CATCHING THE LIGHT

CATCHING
THE
LIGHT

*The Entwined History of
Light and Mind*

Arthur Zajonc

OXFORD UNIVERSITY PRESS
New York   Oxford

Oxford University Press

Oxford   New York
Athens   Auckland   Bangkok   Bombay
Calcutta   Cape Town   Dar es Salaam   Delhi
Florence   Hong Kong   Istanbul   Karachi
Kuala Lumpur   Madras   Madrid   Melbourne
Mexico City   Nairobi   Paris   Singapore
Taipei   Tokyo   Toronto

and associated companies in
Berlin   Ibadan

Library of Congress Cataloging-in-Publication Data
Zajonc, Arthur.
Catching the light : the entwined history of light and mind /
Arthur Zajonc.
p.   cm.
Includes index.
ISBN 0-553-08985-4
ISBN 0-19-509575-8 (Pbk.)
1. Light—History.   2. Light—Philosophy.   I. Title.
QC352.Z35   1993
535'.09—dc20   92-20204

2 4 6 8 10 9 7 5 3 1
Printed in the United States of America

For my wife, Heide

Contents

Contents

I'll tell you how the sun rose a ribbon at a time.
Emily Dickinson

I am the one who openeth his eyes, and there is light;
When his eyes close, darkness falleth.
the Egyptian god Ra, 1300 B.C.

If the light rises in the Sky of the heart . . . and, in the
utterly pure inner man attains the brightness of the sun
or of many suns . . . then his heart is nothing but light, his
subtle body is light, his material covering is light, his
hearing, his sight, his hand, his exterior, his interior, are
nothing but light.
Najm Razi, 1256

All the fifty years of conscious brooding have brought me
no closer to the answer to the question, "What are light
quanta?" Of course today every rascal thinks he knows
the answer, but he is deluding himself.
Albert Einstein, 1951

CHAPTER I

Entwined Lights:
The Lights of Nature
and of Mind

Use the light that is within you to regain your
natural clearness of sight.[1]

Lao-tzu

In 1910, the surgeons Moreau and LePrince wrote about their successful operation on an eight-year-old boy who had been blind since birth because of cataracts.[2] Following the operation, they were anxious to discover how well the child could see. When the boy's eyes were healed, they removed the bandages. Waving a hand in front of the child's physically perfect eyes, they asked him what he saw. He replied weakly, "I don't know." "Don't you see it moving?" they asked. "I don't know" was his only reply. The boy's eyes were clearly not following the slowly moving hand. What he saw was only a varying brightness in front of him. He was then allowed to touch the hand as it began to move; he cried out in a voice of triumph: "It's moving!" He could feel it move, and even, as he said, "hear it move," but he still needed laboriously to learn to *see* it move. Light and eyes were not enough to grant him sight. Passing through the now-clear black pupil of the child's eye, that first light called forth no echoing image from within. The child's sight began as a hollow, silent, dark, and frightening kind of seeing. The light of day beckoned, but no light of mind replied within the boy's anxious, open eyes.

The lights of nature and of mind entwine within the eye and call forth vision. Yet separately, each light is mysterious and dark. Even the brightest light can escape our sight.

—

AS PART OF what I call "Project Eureka," a friend and I have designed and constructed a science exhibit in which one views a region of space filled with light. It is a simple but startling demonstration that uses only a carefully fabricated box and a powerful projector whose light shines directly into it. We have taken special care to ensure that light does not illuminate any interior objects or surfaces in the box. Within the box, there is only pure light, and lots of it. The question is: What does one see? How does light look when left *entirely* to itself?

Approaching the exhibit, I turn on the projector, whose bulb and lenses can be seen through a Plexiglas panel. The projector sends a brilliant light through optical elements into the box beside it. Moving over to a view port, I look into the box and at the light within. What do I see? Absolute darkness! I see nothing but the blackness of empty space.

On the outside of the box is a handle connected to a wand that can move into and out of the box's interior. Pulling the handle, the wand flashes through the dark space before me and I see the wand brilliantly lit on one side. The space clearly is not empty but filled with light. Yet without an object on which the light can fall, one sees only darkness. Light itself is always invisible. We see only things, only objects, not light.

The exhibit reminds me of a conversation I had over dinner with the Apollo astronaut Rusty Schweickart. I asked him about his space walk, specifically about what he saw when looking out into the sunlit emptiness of outer space. He replied that although it was difficult to keep the brightly lit spacecraft and other hardware out of view, if you could do so, then you saw only the dark depths of deep space studded with the light of countless stars. The sun's light, although present everywhere, fell on nothing and so nothing was seen. Only darkness.

Darkness Within

Two lights brighten our world. One is provided by the sun, but another answers to it—the light of the eye. Only through their entwining do we see; lacking either, we are blind.

Arguably the best-studied case of recovery from congenital blindness is the case of S.B., researched by the psychologists Gregory and Wallace.[3] On December 9, 1958, and January 1, 1959, a blind British male, fifty years old, received cornea transplants. For the first time since he was ten months old, he had complete functional use of his eyes. What did he see?

The guardians of S.B. had enrolled him at age nine in the Birmingham Blind School, where he learned the trade of boot repairing. Earning his living by that means, he lived a life that was unusually independent for a blind adult, for example going for long bicycle rides by holding on to the shoulder of a friend. He enjoyed gardening and especially any kind of work with his hands, and was a confident, cheerful, and clearly intelligent man.

Examining him about a month after his operations, Gregory and Wallace asked him about his first visual experience following the operation. S.B. responded by saying that he had heard a voice, the voice of his surgeon, coming from in front of him and to one side. Turning toward the sound, he saw a "blur." S.B. was unsure what the blur was but reasoned that since he had heard the voice of his physician, and knowing that voices come from faces, the blur in front of him must be his physician's face. Faces, even long after the operation, were "never easy," S.B. reported. Nor were his struggles with seeing confined to faces. Gregory and Wallace's research with S.B. (and similar research before and since) has made it clear that learning to see as an adult is not easy at all.

After his release from the hospital, Gregory and Wallace took S.B. to a museum of technology and science. S.B. had a long-standing interest in tools and was clearly excited at the prospect of seeing what he had until now only handled or heard described. They took him to a fine screw-cutting lathe and asked him to tell them what stood before him. Obviously upset, S.B. could say nothing. He complained that he could not see the metal being worked. Then he was brought closer and allowed to touch the lathe. He ran his hands eagerly over the lathe with his eyes shut tight. Then he stood back a little and, opening his eyes, declared, "Now that I've felt it I can see it."

In the case of S.B., the slow process of learning to see continued for the next two years, until his death. Its slow pace and limited success were sources of deep disappointment to him, as they invariably are to all such patients. In many situations S.B., like others, came to neglect sight entirely, for example by leaving the lights off at home and navigating in his accustomed way as a blind man. In most instances the effort to see was simply too great. Those newly given sight may give up completely, sometimes even tragically ending their struggle to see by taking their own life.

In his systematic study of sixty-six case histories of the recovery of sight in those born blind, M. von Senden concluded that innumerable and extraordinary difficulties need to be overcome in learning to see. The world does not appear to the patient as filled with the gifts of intelligible light, color, and shape upon awakening from surgery. The project of learning to see inevitably leads to a psychological crisis in the life of the patient, one that can end with the rejection of sight. New impressions threaten the security of a world previously built upon the sensations of touch and hearing. Some decide it is better to be blind in their own world than sighted in an alien one.[4]

During the last few decades, studies of recovery from congenital blindness have been corroborated and extended through vision research on animals. It is now clear, for example, that if a cat is unable to see forms during the critical period between the fourth week and the fourth month, even if its environment is still light, then the cat will be blind forever. The optically healthy organ of the eye alone is insufficient for sight. During the first months of life, patterns are elaborated in the eye or brain of the kitten by the act of seeing. Without the nourishment of seeing in the first months, these structures decay or are never developed. After the fourth month, the damage is irremediable.[5]

The natural development of human vision is very similar. During a critical window in the first years of life, visual as well as many other sensory and motor skills such as speech and walking are formed. If this opportunity is missed, trying to make up for it at a later time is enormously difficult and mostly unsuccessful.

In the case of Dr. Moreau's eight-year-old patient, after the surgeon had worked with the child for some months, the parents forced him to give the child over to a welfare establishment. By the following year all that the child had learned to see under Dr. Moreau's care was lost. Moreau's account contains a note of exhaustion and regret that even with his full attention, he had accomplished little of permanence. The sober truth remains that vision requires far more than a functioning physical organ. Without an inner light, without a formative visual imagination, we are blind. Moreau writes:

It would be an error to suppose that a patient whose sight has been restored to him by surgical intervention can thereafter see the external world. The eyes have certainly obtained the power to see, but the employment of this power, which as a whole constitutes the act of seeing, still has to be acquired from the very beginning. The operation itself has no more value than that of preparing the eyes to see; education is the most important factor. . . . To give back sight to a congenitally blind person is more the work of an educator than of a surgeon.[6]

Moreau's child held to those modes of knowing that were familiar and reassuring to him: touch, hearing, smell. To do otherwise, to see, would have required a superhuman effort. In many ways we act like Moreau's child. The cognitive capacities we now possess define our world, give it substance and meaning. The prospect of growth is as much a prospect of loss, and threat to security, as a bounty. One must die in order to become. Newly won capacities place us in a tumult of new psychic phenomena, and we become like Odysseus shipwrecked in a stormy sea. Like him we cling tenaciously to the shattered keel of the ship we originally set out upon, our only and last connection to a familiar reality. Why give it up? Do we have the strength to leave, to change? Perhaps the voices encouraging us to venture out on our own belong only to the cruel Sirens? So we close our eyes, and hold to what we know.

Besides an outer light and eye, sight requires an "inner light," one whose luminance complements the familiar outer light and transforms raw sensation into meaningful perception. The light of the mind must flow into and marry with the light of nature to bring forth a world. This urges on us a second inquiry. Having introduced the light of the mind, what, in fact, is the light of nature?

The Darkness That Is Light

My "box of light" exhibit incites in the viewer the puzzling question: what is the nature of this invisible thing called light whose presence calls everything into view—excepting itself? Over time we as a civilization have given many answers. We have called it by the names of gods, or made it an action or attribute of the divine. Even when Western science gave it a more substantial nature, it always reflected our wonder and our capacity to imagine. Early in the seventeenth century, Francis Bacon marveled that the "form and origin of light" had been so little investigated.[7] Why had the specific nature of something so important as light not been discovered? Almost four hundred years later, we, like Bacon, are still naturally curious about what light is made of, how big it is, how it moves, and so on. In other words, we wish to know its physical nature.

In my own professional life, I first sought to understand light by means of laboratory research in quantum optics. In laser experiments performed at institutes in Boulder, Amherst, Paris, Hanover, and Munich, I studied light and the way it touches matter. The more I learned of the quantum theory of light, theoretically and experimentally, the more wonderful light seemed. Even armed with such sophisticated theories, I have no sense of closure regarding our knowledge of light. Far from it, light remains as fundamentally mysterious as ever. In fact, quantum theory has taken the simplistic, mechanistic conceptions of light provided by early science and, on the firm basis of experiment, shown them all to be impossible. In their place, it has framed a new theory of light that every great modern physicist from Albert Einstein to Richard Feynman has struggled to understand—unsuccessfully, as they realized themselves.

Once I understood that for all the power, precision, and beauty of quantum optics, we still do not know what light is, I got excited. The old scientific idols of light, like outdated effigies, have been destroyed; and every attempt to fashion new ones has failed. Our technical mastery of light has opened all the doors once closed by a hasty scientific arrogance. I could not resist entering all the passageways, old and new, within the many-roomed mansion of light. This book is the story of what I found there.

The first thing I discovered was that light has gathered around it innumerable artistic and religious associations of extraordinary beauty. It has been treated scientifically by physicists, symbolically by religious thinkers, and practically by artists and technicians. Each gives voice to a part of our experience of light. When heard together, all speak of one thing whose nature and meaning has been the object of human attention and veneration for millennia. During the last three centuries, the artistic and religious dimensions of light have been kept severely apart from its scientific study. I feel the time has come to welcome them back, and to craft a fuller image of light than any one discipline can offer.

Light touches all aspects of our being, revealing a part of itself in each encounter. A history of those meetings can lead us around and into the nature of light. Long before it was the object of scientific study, light, and especially the sources of light, were venerated as divine—an image of godly nature. Mythologies of all civilizations are rich with tales of the sun, moon, and stars, of fire, rainbow, and aurora. These, too, touch on the being of light, for they are part of the human experience of it. In what follows I will speak equally of the quantum theory of light and the Zoroastrian god of light, Ahura Mazda. I will approach light from many sides, mythic and spiritual as well as historical and technical. Different ages and different peoples have been drawn to one or another of light's many parts. As I studied, it seemed to me that the characteristics of a culture are mirrored in the image of light it has crafted. Each, in its own way, has attempted to uncover light's nature and meanings, and so authored a tale of light. In the telling of that story, the culture reveals as much about itself, about the light of its people's minds, as about nature's light. These twin themes twine their strands around the central axis of these pages like the serpents that spiral the healing staff of Hermes, the god of communication: the changing nature of two lights, the outer light of nature but also the inner light of the mind. I have grown convinced that the two are inseparable.

So, as we follow the historical path of light, we will attend to both the changing ideas of light and the changing human consciousness that studied it. We have looked into the countenance of natural light for many, many years, wondering what or who it is. Light has grown old during the millennia of our looking, its features changing utterly, its tender, childhood face nearly completely disguised. Light displays now a sterner, more useful and mathematical countenance, but even today other visages—artistic, scientific, and spiritual—complement the former. What will light look like tomorrow? Throughout all times and images, the same sun has warmed the earth and lit the planet. From the birth of the very idea of light, through its contemporary maturity, to its final form at the end of time, light will have sighted whole kingdoms and nourished prairie, tree, and flower. How have we changed this thing called light through the lights of our own consciousness? In the mingling of nature and mind arises an understanding of the life of light. This book, therefore, is a biography of that invisible companion who accompanies us inwardly as much as it does outwardly—light.

CHAPTER 2

—

The Gift of Light

Thou sun, of this great world both eye and soul.

Milton

As the beasts of the earth were being created, the Titan Epimetheus (whose name means "afterthought") assumed the task of giving to each some faculty for their protection and survival.[1] To the turtle he gave a hard shell, to the wasp his stinger, to others speed and cunning. When he finally came to the human species, all nature's powers had already been allotted; nothing at all was left for man. In Plato's words, man remained "naked, unshod, unbedded and unarmed." Despondent, the bungling Epimetheus turned to his wise brother Prometheus (whose name means "forethought"). Faced with the helplessness of man, Prometheus daringly stole from Zeus the gift of fire, carrying it to mankind as ancient mariners often carried hot embers, in a giant fennel stalk. From the light of Prometheus' gift, man has kindled his civilizations, his cultures and technologies. The fire and light of Zeus became the possession of mankind.

For his actions on our behalf, Prometheus was cruelly punished by being chained to the mountains of Caucasus, where each day his liver, the seat of life, was torn from his side and eaten by an eagle sent from Zeus. Nor was mankind left to enjoy in peace the gift of Prometheus. Zeus, angered and jealous, commanded the lame craftsman of the gods, Hephaestus, to fashion a seductive automaton, Pandora, whose infamous box Epimetheus greedily accepted. Too late he saw its evil contents. Against his will Pandora lifted the lid and so unleashed on humanity illness, grief, and pain.[2] The gift of fire and all it symbolizes is invariably linked with the burden of care. Under human control, the fire of the gods burns as well as warms, blinds as well as illumines.

—

WESTERN CIVILIZATION BEGAN with the song of a blind bard three thousand years ago who, in singing the *Iliad* and the *Odyssey*, gave voice to the Greek imagination, and so begat Western poetry. Homer's blindness lent his words purity and power. His sense world dark, a godly world rose to take its place, and Homer's memory seemed to reach back to archetypal deeds and an eternal heroic age.

Greek vases show the standing bard rocking as he sang, wrapped in an inner glory, listening, as he spoke, to a voice that sang within him. Like Homer, the old wandering minstrels of north Karelia around the Baltic Sea would rock, eyes closed, sitting upon a log bench, arms locked with those of a peasant, chanting antiphonally their ancient epic, the *Kalevala*.

The Bhagavad-Gita, or "song of god," is the sung response of the minister and charioteer Sanjaya to the questions of the blind king Dhritarashtra. The mightiest earthly power, the king, is blind. He sees through the eyes of another, his charioteer and counselor, whose spiritual gifts extend his sight. When the king asks about the happenings on a distant, sacred field where those he loves prepare to battle one another, Sanjaya is able to see and hear the intimate conversation between the accomplished prince Arjuna and the divine Krishna, who like himself has taken on the form of a charioteer. Here the soul-spiritual faculty of higher sight is separated out, into the person of the charioteer. He becomes the bard who sings to a blind and worldly royalty. The charioteer, like the poet, must see further than others, and speak and steer according to what he sees.

Is it only coincidence that the most renowned soothsayer of antiquity, Tiresias, was blind from his seventh year? He lost his sight for seeing the goddess Athena bathing, that is, for seeing a god unveiled.

The motif is eternal. The light of day makes way for the light of night, of blindness, of inner sight. As Plato wrote: "The mind's eye begins to see clearly when the outer eyes grow dim."[3] The Romantic poet Novalis understood fully the efficacy of darkness. His *Hymns to the Night* begins with a sublime antithesis: "What living, sense-endowed being does not love above all the wondrous appearances spread around him, the glorious light. . . ." And yet Novalis tells us that for all its beauty, he turns away from the day "to the holy, unspeakable, mysterious night." Out of the dark solitude of loss comes the poet's light and voice. In the midst of outer darkness, of blindness, an inner light illumines an imaginal landscape of beauty and reality. The blind bard sings the world he sees into the hearts of his listeners so that they, too, might, for an evening, shed the cares of their world for the beauty of his.

—

WHAT IS THE source of poetic light that illumines the night of Homer's blindness? It is imagination, which is also important to common sight. The light of imagination will occupy half of our history, because of its significance for both the ancient world and poetry and the present world and science. No matter how brilliant the day, if we lack the formative, artistic power of imagination, we become blind, both figuratively and literally. We need a light within as well as daylight without for vision: poetic or scientific, sublime or common.

As we will discover, the mind is subtly and usually unconsciously active in sight, constantly forming and re-forming the world we see. Thus, we participate in sight. The habitual patterns by which we see are set down in the first years of life. Even the simplest, most "objective" acts of cognition require our involvement. In addition, the nature of that participation is specific to each culture and historical period. A Coke bottle dropped from an airplane into a society of bushmen may be seen as many things, but never as a container for a carbonated beverage. Human consciousness has changed over time and differs between cultures.

In antiquity, our role in seeing, in granting meaning to the sense world, was felt more keenly than today; the inner light was closer to consciousness. Unlike the ancient Greeks, we live habitually in a scientific world view that too often treats our participatory role in cognition as unessential or illusory. Yet to see, to hear, to be human requires, even today, our involvement, our ceaseless participation. An example will intensify the argument: the puzzling phenomenon of Greek color vision.

The Wine-Dark Sea of Antiquity

The sun rose on the flawless brimming sea into a sky all brazen—all one brightening for gods immortal and for mortal men on plowlands kind with grain.[4]

Homer, *The Odyssey*

The atmosphere and landscape of Homeric Greece appears at once alike and apart from our own. The sun still rises over plowlands kind with grain, but few of us waken to a bronze sky brightened by gods immortal.

Walking on island shores, a captive of the beautiful nymph Kalypso, Odysseus gazed longingly on the "wine-dark sea" yearning to return to his native Ithaka and his beloved wife,

Penelope. Standing today on an island coastline in the Aegean, I see neither a wine-dark sea nor a brazen sky, but the extraordinary blue of water and heavens I so love.

Of the countless epithets attached by Homer to the sky or sea, none, say linguists, can be understood to mean "blue." The sky may be referred to as "iron" or "bronze," the sea as black, white, gray, purple, or wine-dark—but never blue. Did the ancient Greek lack the experience of blue, was he partially color-blind? Or do we perhaps see here another instance of the presence of an interior light, of the activity of vision? Since 1810, when Goethe first pointed to the curious lack of blue in Greek usage, scholars have been puzzled by it and similar absences in the color language of early Greek poetry.[5]

From the careful analysis of color terms that do exist in ancient Greek, and our modern knowledge of color blindness, compelling arguments have been made against the hypothesis that the Greeks possessed a physical eye different from our own. But then we have shown that sight entails more than an operating physical organ. In considering the following examples of color vision in Homer's Greece, bear in mind the significant inner, psychological pole of sight. If we do so, then we may unravel the puzzle that has confounded so many.

Some five hundred years after Homer, the great student of Aristotle, Theophrastus, wrote a treatise on stones in which he described a stone called *kyanos*, a stone we now identify with the precious blue mineral lapis lazuli. When we encounter *kyanos* in its adjectival form, it is natural to think of it as referring to blue (related to our term "cyan"). Although the association seems natural enough, occurrences in Homer defy that interpretation.

In fury and grief at the loss of his friend Patroklos, Achilles slew Hektor, pierced the heels of this noble son of Priam, and attempted to defile his body by dragging it for twelve days on the plains of Troy. In doing so, "a cloud of dust rose where

Hektor was dragged, his *kyanos* hair was falling about him."[6] Are we to understand that Hektor had a head of blue hair? In order to stay his senseless debasement of a worthy prince and warrior, Zeus sent Iris to Achilles' immortal mother, Thetis, on the sea's floor. Iris, "storm-footed," sprang into the sea and, finding Thetis, bade her join Zeus. Ashamed to mingle with the gods, Thetis "took up her *kyanos* veil, and there is no darker garment," and followed Iris to Olympus.[7] From these and many other instances we learn that *kyanos* meant dark rather than blue. Yet there was no other word for blue in Homeric Greek. Homer and other early poets simply lacked a term for blue. To them blue was not a color in our sense, but the quality of darkness, whether describing hair, clouds, or earth.

Similarly for *chloros,* the term that later Greek color theorists call green, a puzzle confronts us. In the *Iliad,* honey is *chloros;* in the *Odyssey,* so is the nightingale; in Pindar, the dew is *chloros,* and with Euripides, so are tears and blood! From its use, we can see it means not green but moist and fresh—alive. We still speak of unseasoned wood or an untrained laborer as green. For the ancient Greek, these connotations were the primary meanings. So disconnected from the external perception of color were they that the psychological quality of "freshness" or "darkness" could become the perceived attribute. They *saw* the moist freshness of tears and so *saw* green. When enraged, we may remark metaphorically that we "see red." I would suggest that we should understand the Homeric world's use of such color expressions not metaphorically but literally. Neither the sunlight nor their eyes were different from ours. Rather, what they brought as the interpretive light of an antique imagination changed the way they saw, just as a similar light continues to shape the way we see today.

A more recent related example is provided by "The Case of the Colorblind Painter" reported by Oliver Sacks and Robert Wasserman in 1987.[8] Jonathan I. had been a successful painter until, at age sixty-five, he had a minor car accident. He suffered a concussion and the trauma normally associated with such an accident, but took away no lasting physical injury. He did, however, become completely color-blind, a persistent condition whose sudden and unaccountable onset was triggered by the accident. He saw the world, in his own words, as if "viewing a black and white television screen." The case is moving and tragic. An artist whose entire life had been lived through color now saw none. Ophthalmologists and neurologists including Sacks and Wasserman subjected Mr. I. to the full panoply of medical tests to no avail. The cause of his color blindness has remained a mystery. Sacks and Wasserman summarize their study by saying, "Patients such as Mr. I. show us that color is not a given but is only perceived through the grace of an extraordinary complex and specific cerebral process." Moreover, while physiological computations may well be going on, color vision "is infinitely more; it is taken to higher and higher levels, admixed inseparably with all our visual memories, images, desires, expectations, until it becomes an integral part of ourselves, our lifeworld."

The "lifeworld" of Homer on the shores of Troy was profoundly different from our own. His memories, associations, desires, and expectations were unlike those we bring to the battlefield. The mental organ of sight used by the blind bard was then a cultural commonplace, but it was also significantly different from the mind-set with which we see today. We need to soften the notion of ourselves as equipped with fixed vidiconlike eyes and static computerlike brains to produce the equivalent of consciousness. The blossoms of perception unfold out of a far richer and self-reflexive union of mental and natural lights.

In the cases of S.B. and Mr. I., we confront a situation where persons could not see what we would all agree was in fact "there," before their eyes. They lacked an interior light and so remained functionally blind. The opposite situation occurs when someone sees something we would say is "not there." We normally term such experiences hallucinations. They arise when an individual's psychological state is sufficiently strong to fabricate an experience akin to that produced by the senses. Did the ancient Greeks hallucinate the color of their wine-dark sea? The linguistic evidence suggests otherwise, or we would have to imagine an entire culture collectively hallucinating. Yet, in a certain sense, their inner emotions or stage of development did "color" the world they saw. Study of other language groups, for example Chinese or American Indian languages, supports the interpretation that cultures see the world, down to its very textures and colors, in ways significantly different from our own.[9]

Over millennia, the two lights of nature and mind have interacted to present differing worlds to different ages. Like a blind bard granted sight, we will have difficulty at first imagining

Poseidon.

our way into ancient understandings of sunlight and the sighted eye. They will appear, initially, unfamiliar and even absurd. Yet the strangeness may be largely a reflection of the modern imagination we bring to ancient experiences. At every stage, we will need to reimagine the universe, to participate in it empathetically in order to hear the epic song of light.

The magnificent bronze statue of Poseidon, lifted from the Aegean to grace the National Museum of Athens, has only dark hollows where eyes once were. They were not empty in 450 B.C., when special inlaid gems filled those now sightless pools. Here was the sacred seat of a special glory that brought cognitive life to the supple and powerful figure. When toppled from his pedestal and cast into the sea, the gems—Poseidon's eyes—were stolen. The god who once ruled the sea, now blind, was deposed. Our story of light begins with the ancient and sacred understanding of the eye, so close to light. Empedocles, physician and demi-god, will reset the gemstones in Poseidon's face. Later, others will remove them.

The Lantern and the Eye

The light of the body is the eye: if
therefore thine eye be single, thy whole body
shall be full of light. But if thine eye be
evil, thy whole body shall be full of
darkness. If therefore the light that is in
thee be darkness, how great is that darkness!
                                    Matthew 6:22–23

In his book on ancient philosophers, Diogenes Laertius tells the tale of a pestilence that struck the Sicilian city of Selinus in the middle of the fifth century B.C.[10] From the stagnant, sewage-filled waters of the main river there arose disease and death, claiming the lives of "both citizens and women." Hearing of Selinus' distress, the noble physician, scientist, statesman, and poet Empedocles came from the neighboring town of Acragas dressed in the purple robe of wealth, girded with a belt of gold. On his feet he wore bronze slippers, on his head a laurel wreath; behind him followed a train of young boys to attend his needs. Discovering the source of the pestilence, Empedocles caused two nearby rivers to be rechanneled, mixing thereby their sweet, flowing waters with the rank river water of Selinus, and so relieving the Selinuntines from their ills.

The story is entirely believable, considering the appalling state of rivers around urban centers in present times. Paris, perched upon the Ile de la Cité, was famous even in the Roman period for the stench of the Seine; the Baltic Sea, once a magnificent natural resource, has become the poisonous dumping ground for industrial Poland. For Empedocles to have identified the cause of disease, conceived a plan of action, and then engineered and excavated at his own expense a canal suited to dispersing the fetid waters of Selinus is an action deserving of high commendation. So we are not surprised to learn that when Empedocles appeared to the Selinuntines afterward, he was praised and worshiped as a god. What does surprise us is that, in response to their praises, Empedocles threw himself into a fire, apparently without harm, to confirm their opinion of his divinity.

In Empedocles, we find not only an impressive early scientist whose specific ideas on vision will concern us, but also the last, belated example of a human type that vanished from Greece at his remarkable death when he disappeared into the volcanic crater of Mount Etna, bronze slippers and all. Empedocles was not only a scientist-physician but also a poet and shaman who wrote, in addition to his insightful book *On Nature*, a rather more puzzling spiritual-religious tract, *Purifications*.[11] We possess only fragments of these two works, but from them we can still form an opinion of the range and character of Empedocles' person and thought.

In the *Purifications*, we learn from Empedocles himself of his divine origins, that he has been condemned to "wander thrice ten thousand seasons far from the company of the blessed, being born throughout the period into all kinds of mortal shapes."[12] He is a god sentenced to live as a bird, a mortal, and in countless other forms for the horrid transgression of eating the human flesh of sacrifice. As we study his account of sight, remember the paradoxical blend that is Empedocles, shaman and scientist in fifth-century Greece, less than one hundred years before Plato. The world of Homer, of the gods, the pageants of mystery religions, and the guarded rites of initiation are not distant from the natural science of early Greece. A spiritual cosmos provided the protecting chambers in which the birth of natural science took place.

According to Empedocles, the divine Aphrodite, goddess of love, fashioned our eyes out of the four Greek elements of earth, water, air, and fire, fitting them together with rivets of love.[13] Then "as when a man, thinking to make an excursion through the night, prepares a lantern," lighting it at the brightly blazing hearth fire and fitting it around with glass plates to shield it from the winds, so did Aphrodite kindle the fire of the eye at the primal hearth fire of the universe, confining it with tissues in the sphere of the eyeball.[14] Marvelous passages were fitted into the eye, permitting it to transmit a fine interior fire through the water of the eye and out into the world, thereby giving rise to sight. Sight proceeded from the eye to the object seen; the eyes rayed out their own light.

When, a few hundred years after Empedocles, the evangelist Matthew wrote, "the eye is the light of the body," he was thinking not only metaphorically but scientifically. The image of the eye as a lantern was a cultural and scientific commonplace when Matthew was composing his gospel.[15]

In the struggle to unravel the mystery of sight, sunlight played a lesser role. Empedocles recognized the existence of the sun's light; that much is clear from other fragments such as "It is the

Earth that makes night by coming in the way of the [sun's] rays,"[16] an astute observation for the time. He seems, however, to have considered sunlight as only part of the whole process, and recognized that something more was required for vision, something essential provided by man: *the light of the body*.

Platonic Sight

Plato, like Empedocles, was permitted to study the secret doctrines of Pythagoras, at least until he (again like Empedocles) betrayed Pythagoras' teachings to the uninitiated through his writings. Plato's account of vision is, not surprisingly, similar to, if fuller than, that of Empedocles. When blended together with the later geometrical tradition of sight begun by Euclid and the medical tradition codified by Galen, Plato's treatment would persist for almost 1,500 years! In this tradition, the light of the eye played fully as important a role as the light of the sun.

According to Plato, the fire of the eye causes a gentle light to issue from it. This interior light coalesces with the daylight, like to like, forming thereby a single homogeneous body of light. That body, a marriage of inner light and outer, forges a link between the objects of the world and the soul. It becomes the bridge along which the subtle motions of an exterior object may pass, causing the sensation of sight.[17]

In this view, two lights—an inner and outer—come together and act as the mediator between man and a dark, cavernous external world. Once the link of light is formed, the message may pass, like Iris, Homer's messenger goddess, from one world to the other. The eye and the sun display to Plato a deep harmony, one still appreciated by the German poet Goethe when, in the introduction to his own *Theory of Color* (1810), he penned the poem:

Were the eye not of the sun,
How could we behold the light?
If God's might and ours were not as one,
How could His work enchant our sight?[18]

Once again, the mind's eye is not passive, but plays its own significant part in the activity of seeing. The image of an interior ocular fire captured vividly the ancient sense of that action, so convincingly that it dominated philosophy for 1,500 years.

We come to know the world, in large part, through sight. Quite naturally, Plato used sight as a metaphor for all knowing, calling the psyche's own organ of perception the "eye of the soul" or "mind's eye."[19] Our word "theory" has its origin in the Greek word *theoria*, meaning "to behold." To know is to have seen, not passively but actively, through the action of the eye's fire, which reaches out to grasp, and so to apprehend the world. Our activity, present in seeing and knowing, is an element integral to the Platonic understanding of vision. Sight entails the seer in an essential, formative action of image making or imagination. To such as Moreau's child or to S.B., the effort of that constructive act was a constant and exhausting reminder of their past blindness. To us who see, the world is instantly and effortlessly intelligible; at least most of the time.

Consider the figure on page 23. It is but one of many similarly "ambiguous figures." Allow yourself time to play with it. At first only one figure shows itself, an old woman or a young girl. Without an iota's change on the "objective" printed page, the delicate chin of the young girl becomes the lumped nose of an old hag. Feel the shift from one picture to the other. It takes place entirely within you. With a little practice you can even control what you see.

The physical difference between one image and the other is nil, while the "soul distance" between them is huge. What has changed? Your own activity; the character of your participation

Old woman or young maid?

can shape and reshape itself, and you can feel it. With every act of perception, we participate unawares in making a meaningful world. In response to outer light, an interior light flashes, bringing intelligence with it. It is the light that did not brighten the newly opened eye of Moreau's child when turned to see its first light.

Times of Transition

In the Bhagavad-Gita, in Homer, Empedocles, and Plato, vision entails an essential human activity of movement out from the eye into the world. In the centuries following Plato, a shift gradually took place that only reached its conclusion with René Descartes in the seventeenth century. The concerns of science changed during this long period. The influence of Plato and then of Aristotle lingered long into the medieval period. As long as this was the case, sight was as much or more a soul-spiritual process as a physical one. By the sixteenth century, however, a profound shift seems nearly mature. Natural philosophers such as Kepler and, to a much greater extent, Galileo are less concerned with the soul's translation of external stimulation into meaningful perception, and more preoccupied with the physics of the eye viewed as an inanimate, physical instrument. The change is not universal, swift, or uniform, but a watershed is crossed nonetheless, first by those few scientists in the oft-dangerous vanguard of research. In their hands, sight becomes a question of mechanics rather than a species of soul-spiritual activity so characteristic of many earlier thinkers.

The shift is characteristic and of central importance. We meet it first in the evolution of man's experience of seeing. We will discover it again when we study light itself. What begins as a lively, soul-spiritual experience, be it of light or sight, attenuates, clarifies, and divides into optics and psychology. More than an interesting historical observation, our changing view of light is symbolic of a major change in consciousness, an important threshold crossed in the history of the mind.

Like the ambiguous figure, nature presents herself in indefinite guises. How we see her depends as much on us as on her. Only together do meaningful worldly images arise. The watershed crossed, therefore, is not the divide between ignorance and wisdom, but more like the ambiguous shift from young girl to old woman. Therefore as we read the history of science, we must be ever conscious of the individuals who enacted it. Their eyes saw, their hearts yearned for knowledge, and out of their being ways of seeing the world were born, flourished, and died. One way of seeing became for a time the way of many, until a fresher, more congenial view appeared.

The delicate beginnings of the transition to a mechanical conception of seeing were evident by 300 B.C. in the optical studies attributed to the great Alexandrian mathematician Euclid. In his book *Optics* he provided a brilliant geometrical treatment of sight. Euclid continued to believe that a visual ray was primary to the whole process of vision, and advanced several very sensible arguments in favor of the position.

For example, we often do not see things even when looking at them. Drop a needle on the ground, Euclid suggests, and then wonder as you search for it why you don't see it immediately. Your field of view certainly encompasses the needle. In modern terms, the needle is certainly imaged on the retina, but remains unseen. Then suddenly, in a flash, you see it. If sight depends only on light from outside falling on objects, and then traveling into the eye, one would see it immediately. Obviously light was being reflected from the needle and into the eye throughout the search, so, reasoned Euclid, sight cannot in the first place depend on external light. The puzzle is solved, however, if we adopt the doctrine of the visual ray. In searching for the needle, the eye's own visual ray reaches out and passes back and forth across the ground. Only when it strikes the needle do we see it!

The visual ray of Euclid is different in important ways, however, from the luminous and ethereal emanation of Plato and Empedocles. In Euclid's hands, the eye's fiery emanation has become a straight line, a visual ray, susceptible to deductive logic and geometric proof. His extensive mathematical studies yielded many fruits and became the basis for later Arab investigations and for laying the foundation for the discovery of linear perspective by Brunelleschi, Alberti, and Dürer centuries later. But mathematization came at a price. It distanced man from the earlier and more immediate experience reflected in the Platonic understanding of vision.

The significance of mathematization should not be underestimated. Without abstraction, science as we know it cannot exist. Yet in order to analyze one must stop experiencing and go on to represent the object of study with thoughts of crystalline clarity, for example, with mathematical concepts. Euclid did just this. Plato's somewhat elusive, immaterial bridge of light between object and eye, became through Euclid a geometry of visual rays, cones, and angular measurement. Everything needed for the study of geometrical optics was developed, but in the process one can detect an important distancing from the subjective human experience of seeing. Euclid's meticulous mathematical style of argumentation has replaced the more poetic treatment of Empedocles or Plato. As every physicist knows, the elegant forms of mathematics can easily outshine the dull stirrings of experience, and eventually come to replace the phenomena they originally were invented to describe. Euclid's handling of light foreshadows the growing separation of sight as lived experience from sight as a formal object of investigation. The history of light has turned a corner, and with it the mystery of sight entered a new phase, one that blossomed first in Arab lands, to culminate finally in the work of another great geometer and mathematician, René Descartes.

The Arab Connection

Toward the end of the Roman empire the stage was set for further developments in the history of the mind. The closing of the Platonic Academy in A.D. 529 by Justinian was the final death knell of Greek philosophy in the West and the dawn of the Dark Ages. For many centuries, the Academy had been a sanctuary in which the ideas of Plato and his followers flourished. With the rise of Christianity, however, pagan thought was in danger of being eradicated. In A.D. 389, the great library in Alexandria with its half-million scrolls was destroyed by rioting Christians. Under a state that sanctioned the Roman Church, the Platonists, who still revered the pagan gods, were persecuted and hounded as dangerous heretics. When Justinian's soldiers swept into the Platonic Academy, the last disciples of Plato had to flee Athens. The seven great sages of the Academy departed with their precious books bound for Persia where the emperor Khurso I received them graciously at his magnificent summer palace in Jundishapur (near what is today Dizful, Iran).[20]

In the court of Khurso I, and at the illustrious Academy of Jundishapur, literature, the arts, science, and philosophy flourished. The Athenian refugees found here a cosmopolitan atmosphere of remarkable tolerance. The indigenous religions of Zoroastrianism and Manichaeism mingled with eastern religious thought, as well as pagan, Christian, and Jewish influences. Jundishapur was founded as a prisoners' camp following the defeat of the Roman emperor Valerian in A.D. 260 by Shapur I. By the sixth century it had become the greatest center of learning in the world, boasting an outstanding astronomical observatory, medical school, and the world's first hospital. Jundishapur was known then, and for centuries thereafter, for its physicians and wise counselors. The rise of Islam blunted the impact of Jundishapur, but the leaders of Jundishapur's Academy were the nucleus around which the scholarship and learning of Islam formed.

With the rise of Islam in the seventh century, a cultural revolution of unprecedented scope took place on the Arabian peninsula. Following the establishment of the new religion by Mohammed and a system of governance for the vast empire won through holy wars, Islamic scholars became tremendously active in collecting and translating Greek manuscripts. Baghdad, during the ninth century under the guidance of the scholar and translator Hunayn ibn Ishaq, became a great center of learning, and Arab science and scientists rose quickly in importance. While thinkers in the West forsook the concerns of Hellenism for religious questions, especially the matter of salvation, philosophers and physicians of the Islamic Near East, under the influence of Jundishapur, were busy mastering, commenting on, and furthering the knowledge of antiquity.

The famous philosopher, mathematician, astronomer, and optician Ibn al-Haytham figured prominently in these developments.[21] In his hands, the history of sight took another significant step away from earlier and more spiritual or psychological views and toward a mathematical and physical theory of vision.

Born in Basra (Iraq) in A.D. 965, Ibn al-Haytham, or Alhazen as he came to be known in the West, became the greatest optical scientist of his age. As a child and young man, Alhazen had attempted to attain knowledge of truth through the Islamic religious sciences of his day. Dismayed by the elusiveness of this goal and the rancor he saw between competing religious sects, he resolved to concern himself with a "doctrine whose matter was sensible and whose form was rational."[22] Truth was one, he felt, and throughout the following decades he maintained his initial resolve to avoid the vagaries of the spiritual sciences. Instead, he produced dozens of treatises on mathematical and scientific subjects, the most influential of which was his *Optics*. One hundred and fifty years after his death in 1040, the *Optics* was translated into Latin and subsequently became the foundation for future optical research. Two aspects of his work will be of special concern to us: his replacement of the Platonic theory of vision with his own quite different theory, and his study of the *camera obscura*. Both reflect Alhazen's reimagination of light.

—

THE PROMINENT GREEK accounts of vision had given full weight to the inner activity of the seer. As we have seen, this came to be embodied in their view that a pure fire, essential to sight, resided within the eye and rayed out, sunlike, to illuminate the world. This view was taught in various forms in the West until the twelfth century, for example by the great teacher William of Conches at the cathedral schools of Chartres and Paris. A profound student of Plato, Conches also drew from Galen the view that food was transformed from matter into a spiritual light in a series of stages. Its first transformation occurred in the liver, where it became "natural virtue." Passing then on through the heart, it became "spiritual virtue," moving finally into the brain, where it was refined into a luminous wind that animated the organs of sense and provided the interior ray of the eye.[23]

Another important Greek school of thinking held that vision occurred by the transmission of husks or forms (called *eidola* or *simulacra*) from the object to the eye. The atomists of the Greek world believed that films or images peeled off objects, or were impressed onto the air by them, and streamed to the observer, where they entered the eye. The tiny reflected image of the world that is visible when we look at the dark pupil of our neighbor's eye was taken by them as evidence of these husks. Obvious problems existed with this theory. How, for example, did a husk the size of a mountain become small enough to enter the eye?[24] This view likewise found its reflection in the Middle Ages, but setting it aside for the moment, we can return to the Arab connection.

In the Near East, a view of vision was being developed that was complementary to Plato's view as taught at Chartres. It emphasized outer light, and received a special impetus in the Arab world. Alhazen marshaled a series of logical arguments in support of the view that sight proceeded not partly, but entirely by means of light entering the eye from objects around us. He considered the following situation. One cannot long look at the sun without great pain. If one maintained a theory in which the flow was away from the eye, then how could there be pain? If, however, there were some kind of transmission from the sun to the eye, its overwhelming action on the eye could account for the discomfort. Another of his arguments concerned afterimages.

Stare at a bright light or a window for thirty seconds and then close your eyes. A clear sense impression floats in view with the same contours as the original but usually showing colors complementary to those of the lights seen. Again Alhazen took this as evidence that something affects the eyes from without, impressing itself on the eye so strongly that an effect persists even after the light is extinguished. These, and many other phenomena, were joined with carefully reasoned arguments to reject the visual theories of Plato and others. Alhazen conceded that mathematicians may still find it useful to draw "visual rays" from the eye to the object for the geometrical study of light, but in doing so they "use nothing in their demonstrations except imaginary lines . . . and the belief of those who suppose that something [really] issues from the eye is false."[25]

Thus have the rays of Empedocles' interior fire been quenched. In their place Alhazen offers a carefully elaborated theory of exterior, physical rays that can be combined with Euclid's precise language of mathematics to provide a compelling scientific account of vision. The eye, once the site of a sunlike, divine fire, fast became a darkened chamber, awaiting an external force to enlighten it.

Vision in a Dark Chamber

At this time, a device known as the *camera obscura*, which literally means "darkened chamber," affected the scientific imagination so greatly that by the seventeenth century it had become *the* model for the eye. While antecedents can be found earlier, its first clear description appears in the writings of Alhazen.[26]

On a bright day, stand within a darkened room. Punch a small hole the size of this letter "o" in an opaque curtain over a window in the room. Outside is a bright world, inside a dark room; they are connected only by the light filtering through a single small aperture. On the wall of the darkened chamber opposite the hole, a wonderful inverted image of the outside scene appears in full detail. In his study of the *camera obscura*, Alhazen arranged several candles in a row on one side, their flickering images then appearing in a similar but now inverted row on the screen. If one held a theory of "husks," they all must pass through the same small aperture without interfering with one another. Somehow the light from each candle passes simultaneously through the same single point without obscuring the image. Amazingly, an entire landscape, rich in color and detail, can make its way undisturbed into a *camera obscura* through a single tiny hole. In fact, make the hole too large and the image, though brighter, becomes blurred.

The specific connection of this experiment with vision had to wait four hundred years, until the Renaissance genius Leonardo da Vinci made the extraordinary suggestion that the eye itself is a *camera obscura*. The eye, too, according to da Vinci, is a dark chamber into which an image of the world is projected.

In the first years of the seventeenth century, the mathematician and astronomer Johannes Kepler developed a complete geometrical explanation for the *camera obscura*, and went on to give a detailed and successful explanation of the optics of the eye and vision in terms of it. As in the *camera obscura*, the outer world was projected onto an inner screen in the eye. Kepler declared that "vision occurs when the image of the whole hemisphere of the world that is before the eye . . . is fixed on the reddish white concave surface of the retina."[27] Yet Kepler, like so many scientists before him, was deeply disturbed by one fact: the image on the screen of a *camera obscura* is inverted! How can the image on the retina be upside down when we see the world right side up? Innumerable fantastic schemes had been invented to rectify the image, but Kepler's geometrical arguments were so tightly reasoned that, even in the absence of direct observational evidence, there was simply no escaping the conclusion: the retinal image must be inverted. Kepler accepted this and left to others to explain how the image could be righted.

Today we locate the solution to this problem in the mind, associating it with psychology of vision. In Kepler's words,

> How the image or picture is composed by the visual spirits that reside in the retina and the optic nerve, and whether it is made to appear before the soul or the tribunal of the visual faculty by a spirit within the hollows of the brain, or whether by the visual faculty . . . all this I leave to be disputed by the physicists [philosophers]. For the armament of opticians does not take them beyond this first opaque wall encountered within the eye.[28]

At this point optics ends and the light of the body, that is the soul's activities, must be engaged in order for us to see the world right side up.

Descartes

Experimental verification of Kepler's deductions was finally achieved by René Descartes. Descartes's optical studies contain a revealing illustration of the visual system that unwittingly depicts not only Descartes's understanding of the anatomy of the eye and of visual optics, but also his philosophy of perception.[29]

Above and at some distance from an enormous eye are arranged three geometrical objects: a circle, a diamond, and a triangle. Rays are drawn from them through the lens of the eye below, and are focused on the retina. The rear membranes of the eye have been removed in order that the philosopher (Descartes himself?) can view the three images projected onto the rear surface. The external world depicted in the upper part of the picture is shown in light; the lower part surrounding the observer is dark. As with Alhazen, for Descartes the world is bright, and the eye dark.

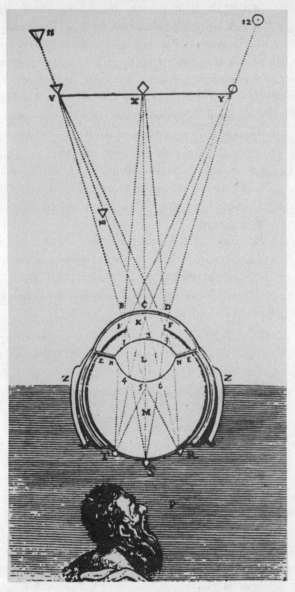

Optical analysis by Descartes. The philosopher viewed the world through an ox's eye whose back had been scraped to make it transparent. The image he saw was inverted.

The illuminating interior ray or fire had vanished. Yet Descartes still holds a two-stage theory of vision. In the first, light (which he conceived of as material and mechanical) is conveyed through the physical organ of sight to a common sensorium in the body. The mechanical stimuli are then, in Descartes's view, "perceived" by a spiritual principle within man. For Descartes the world of extension, of substance—*res extensa*—reached all the way into the body but could not of itself complete the process of vision. A spiritual principle, the mind or soul—*res cogitans*—was still required. Like the philosopher in the illustration observing the flickering retinal images from his dark vantage point, the immaterial mind observed the mechanical proddings of the world in the sensorium.

Although the light of the eye that reached out and granted meaning to raw sensation had retreated from the body, it remained in Descartes's dualist position as a disembodied spirit, a vestige of the past. Yet even this faint echo of a Greek heritage was destined for at least temporary extinction.

Modern Sense Physiology

We shall, sooner or later, arrive at a mechanical equivalent of consciousness.

Thomas Huxley

The final development in the evolutionary drama of sight (and I must leave much out) occurred in our century. By the mid-1900s, the neurophysiology and psychology of vision had advanced to an extraordinary degree. The detailed knowledge we now possess of brain structure and function, of the neural anatomy of the eye and visual pathways, is truly staggering. In the flush of excitement that naturally accompanies a century of discovery, many feel that they hold now in their hands "the mechanical equivalent of consciousness," as Thomas Huxley called it.

The Harvard biologist and Nobel laureate David Hubel speaks for many scientists when he states that the brain is a machine "that does tasks in a way that is consonant with the laws of physics, an object that we can ultimately understand in the same way we understand a printing press."[30] Moreover, contrary to Descartes, we have no need to appeal "to mystical life forces— or to the mind" to account for perception, thought, or emotion. They are purely and simply states of the physical brain.

Hubel rightly recognizes the profound implications of this view for everything we do. Our image of the mind sets the agenda for everything from education to love relationships. According to Hubel, once we understand that mind is an illusion, and that brain is the sole reality, then we can restructure our systems of education and social institutions to serve the brain, not an antiquated notion of "spiritual man."

In traditional language, the substitution of a purely material and sensual image for a spiritual reality is idolatry. In his insightful little book *Saving the Appearances*, Owen Barfield suggests a connection between the biblical injunction against idolatry and the veneration of models so common to modern scientific practice.[31] Scientific models certainly have their rightful place. But when does a model become an idol, that is, when is it taken for something other than a model, becoming "reality"? The model of an atom as a miniature planetary system is helpful only as long as it is not taken literally. Quantum physicists discovered long ago the dangers of idolatry. Neurophysiologists have yet to learn the lesson. For many of them, the brain has become an idol; it has become quintessential man.

The dangers associated with this kind of adulation of the brain are innumerable. The image we have of ourselves is a powerful thing; it shapes our actions, and so also the world we fashion for us and for our children. It is important, therefore, patiently and carefully to distinguish between idol and fact.

I am not suggesting a simplistic Romantic return to the past. There is no turning back. Yet are Hubel and the legions of scientists who think like him right to reduce our humanity to brain function? The answer is, quite simply, no. The brain as described by Hubel is a carefully crafted and dazzling image fashioned from the fruits of scientific research, one full of insights, but which is ultimately mistaken for something it is not. Is it possible to embrace the results of science without falling into such idolatry? Yes, but this, perhaps more than any other, is the challenge we confront in our times. Our success or failure in fashioning a nonidolatrous science will determine much for our future.

Rekindling the Fire of the Eye

The movements we have traced are like a contrapuntal harmony where one melody sounds against the other. As the light of the eye dims, that of the world brightens. As the beacon of the eye gradually retreats, the power of sunlight projects itself deeper and deeper into the human being until finally the ethereal emanations of Plato, and even the Cartesian spectator, vanish from the Western scientific sense of self. Yet some data and scientific developments indicate the possibility of a "postmodern" view of self and vision that has room in it for the light of the eye. In them, the interior ray may once again find a place, even if under another guise.

We have learned that our consciousness is not immutable. Our habits of thought become perceptions, and while powerful and pervasive, these are not universal or "true." We should learn to take responsibility for them. Do they accord with our deepest intentions and the good of our society and planet? Or do we need to "reimagine" ourselves and our world? In this way Matthew's remarks make real sense: "If thine eye be evil, thy whole body shall be full of darkness. If therefore the light that is in thee be darkness, how great *is* that darkness!" Our light, a light of meaning, fashions a world, forms it from the light of day. If our light be darkness—be evil—then we bring darkness and evil into our whole body, personal and social. If it be light—be good—then health flows into us, and into the world.

Plato's light of the eye was a light of interpretation, of "intentionality," as modern phenomenologists would say; a light that grants meaning. Cognition entails two actions: the world presents itself, but we must "re-present" it. We bring ourselves, with all our faculties and limitations, to the world's presentation in order to give form, figure, and meaning to that content. The beautiful and productive images we craft on the basis of experience are images only—fruits of the imagination. They are no less true for being so. If in our enthusiasm we forget this, then images become idols demanding propitiation at their high altars. Nor should these reflections become grounds for abandoning the path of knowledge, because it is a philosophy of growth and development. The organs of insight we bring are neither fixed nor limited, but malleable and expansive. Thus the importance of integrating the insights into light gained by artistic and spiritual disciplines as well as scientific ones.

CHAPTER 3

Light Divided: Divine Light and Optical Science

Our original question—what is the nature of light?—has been answered differently by different peoples. To the Egyptians it asked after man's relationship to the god Ra. They sought first a moral or spiritual answer, not a mechanistic one. By contrast, we search to explain the nature of light by tracing light rays through intricate optical systems. We seek light's mathematical and physical lawfulness. The sequence of words—what is light?—does not have a unique meaning. The Egyptian answer is utterly different from that of quantum optics, but are they necessarily at odds? Or does the Egyptian yearn to know a different part of light's expansive being?

We should be open to the possibility that the essential questions posed about nature in past or future ages may be quite different from those posed by us today. As C. S. Lewis wrote in his lovely book *The Discarded Image*, it is not that our present-day understanding is unsubstantiated, but we should realize that "nature gives most of her evidence in answer to the questions we ask her."[1] The questions we ask, as well as the answers we are willing to accept, reflect our temper of mind. The images of one age may be discarded by another less because of new discoveries than because of new priorities and new questions, all of which reflect a changing psyche.

By remaining open to conflicting proposals regarding the nature of light, we allow the wide-ranging drama of the past its place, and so read the full biography of light rather than only the fragment we have written ourselves. We can also wonder afresh at our present understandings of light and see the future as being still undetermined. As we approach the ancients' understanding of light, we should leave our own hard-won, contemporary images at the threshold to see as others have seen.

The Lost Eye

I am the one who openeth his eyes, and there is light; When his eyes close, darkness falleth.

> Ra speaking; from the
> Turin papyrus, 1300 B.C.

Two eyes looked down on the civilization of the Nile, the "two eyes of Horus," the sun and moon. No more significant symbol existed in ancient Egypt than the eye of the sun-god Ra. His eye—the sun—was creative, his vision was life itself. It was said that mankind arose from the tears of his eye. In Egyptian, the very words for tears and men sounded similar.

The nature of light was clear to the Egyptians. As the priest-scribe wrote in the above fragment 3,300 years ago, when Ra "openeth his eyes . . . there is light; When his eyes close, darkness falleth." The gaze of Ra was the light of day. For men and women of that civilization, to stand within daylight was to stand in the sight of their sun-god. The power of vision to illuminate the world was universalized, projected onto the grandest scale, becoming the brightness of day. The gaze of God was light. *Light was God seeing.*

We are reminded of the Greeks, who felt the force of their vision, the "light" of their own eye, and developed a theory of

The solar eye of Egypt, taken from a funerary papyrus, 1000 B.C.

vision based, in part, on that experience. From the mythology of the Egyptian world with its multitude of stories about the eyes of Horus or Ra, we come to realize that prior to the individualized light of Greek visual theory, sunlight itself was felt to be an emanation of an eye, that of the sun-god Ra. In neither case was light a substance or thing, but rather was felt to be the power of seeing. To see was to illumine. For Empedocles, the human eye was like a lantern lit at the hearth of creation. When open, it rayed forth into the world and man saw. For the Egyptian priest, the sun itself was an eye, which, when open, brought the day and, when closed, the night. The kinship between the eye and the sun was felt deeply for many centuries, from ancient Egypt to the medieval mystics. In Persian and Greek mythology the identical image reappears—the sun and moon are the eyes of gods placed into the heavens.[2]

The earliest answer given to our question—what is the nature of light?—must be: it is the sight of God. Mankind, formed from the tears of Ra or by means of his sight alone, shares somewhat in his nature, like debased gods.[3] By the time of the Greek philosophers, we, like the gods, illuminate the world with our sight. The sight of Ra lights up the cosmos; the sight of man lights up our personal world.

The change from universal illumination and the eye of god to the human is beautifully told in the Egyptian story of the "lost eye" of Horus or Ra that appears in many variations within Egyptian mythology.[4] It seems that the Eye of the supreme god of Egypt wandered away from its appointed course as the Sun, becoming lost in the watery depths of this world and living as a lioness in the eastern mountains of the sunrise. Ra sent Shu and Tefnut in search of it, but by the time the Eye was found and returned to the face of Ra, another eye had been fashioned to take its place. The original Eye was outraged, but the god Thoth pacified and healed it. Ra went still further and made a place for it within the enclosed serpent form of the *uraeus*, and placed it in the middle of his forehead "where it could rule the whole world." In the depiction of the pharaohs, this same emblem rested on their heads. The Eye of Ra, the Sun, was no longer free to rove unfettered, but was now forever encompassed by the asp-serpent. Delimited and individualized by the power of the serpent, it became the ruler of *this* world. The all-mighty pharaoh, therefore, is depicted as crowned with the *uraeus*. The eye of Ra becomes the eye of man; the light of god the light of man.

Light was, and has remained, an aspect of God. Imaged in countless ways as sight or angel or one of a thousand other things, it has been inseparable from man's groping to represent the spirit. From Egypt we move to ancient Persia, where light, darkness, and the divine unite to form a glorious religious universe.

Light in a Dark World

Legend has it that at the age of thirty, Zoroaster, whose life had been spent in careful attention to the path of righteousness, stood in the river Daiti in order to draw water for ritual libations.

Drawing the water, he looked up and saw a man fair, bright, and radiant, whose silken gown was woven of angelic light. The divine emissary led him to the great god of creation, the god of light, Ahura Mazda, for whom Zoroaster had longed constantly. So began the visions and ministry of the Persian prophet Zoroaster in the second millennium B.C.[5] The Egyptian imagination had elucidated the kinship between light and eye in god and man. The Persians, by contrast, explored the origin and signification of light—and its negation, darkness—through a majestic cosmic myth of creation.

In the dualistic religion of Zoroastrianism, the antagonism between light and darkness was embodied in the form of warring spiritual powers. Ahura Mazda, the uncreated creator of all that is good on heaven and Earth, was opposed by an adversary, also uncreated, but wholly malign, the "Hostile Spirit," Angra Mainyu, or Ahriman.[6] The ultimate polarity of good and evil, of light and darkness, is expressed in the struggle between these two beings and their company of created spirits. Ahura Mazda fashioned the world first in a purely spiritual, disembodied state beyond the reach of Angra Mainyu. Following this came a second creation of the world, but now in a material existence, which Angra Mainyu together with his Daevas (demons) immediately attacked, mingling his own darkness with the static, luminous perfection of the original creation. As a consequence, the seas became salty, fire was tainted by smoke, deserts appeared, and all physical existence became a mixture of good and evil, light and darkness.

The earliest earthly source of light, fire, had always played an important role in the practice of worship of Indo-Iranian civilization. Under the influence of Zoroastrianism, it gradually grew to assume ever greater significance in temple and home rituals, so that by the sixth century B.C. there were great numbers of fire temples. Their holy fires were ritually fed in a threefold manner with dry fuel, incense, and animal fat by priests who never allowed them to fail. Perhaps worshipers saw in the fires a symbol for the redemption of the earth. The wood, itself a mixture of light and darkness, was transmuted by the alchemical action of fire, as if by the fiery might of Ahura Mazda, to become radiant light once again. The smoke and residual ash were the irredeemable darkness woven into creation by Angra Mainyu and now cast out by the fire of god.

The mission of human beings, as a creation of Ahura Mazda, was also symbolized in the fire rituals. Ahura Mazda and his Yazatas (beneficent spirits) shared the task of redeeming the world by separating good from evil, the light from the darkness, with humans.

According to Zoroaster, we live in a time of suffering, sorrow, illness, and death, in a time when the worlds of light and darkness are mingled. Yet he insisted that we are not to flee the suffering of the world but should assume the responsibilities entrusted to us by Ahura Mazda. A third age will come when the pristine first creation is restored, after the ultimate defeat of Angra Mainyu and his legions by Ahura Mazda, the Yazatas, and man. Then good will be separated from evil, the light from the darkness, and redemption accomplished. We now live in the middle of the "Three Times" that frame Zoroastrian cosmogony—Creation, Mixture, Separation. The role Zoroaster assigns to man in world evolution is one of enormous dignity and responsibility for the future, and one that places him squarely in a cultic imagination whose root metaphors are light and darkness.

Zoroaster's spiritual universe was peopled with innumerable divine beings whose primeval actions had given rise to all existence. Light and darkness surround us as images of those ancient deeds. As creations of light, human moral actions now will determine the future forms of existence for better or worse. Our spiritual actions, like those of the gods, will one day image themselves in light or darkness. What is inner will become outer.

Angelic Light—Human Light

How you have fallen from heaven, bright son of the morning, felled to the earth ...

Isaiah 14:12–15

On the first day God created light; on the fourth He "made two great lights, the greater to govern the day and the lesser to govern the night; and with them he made the stars." Placed in the vault of heaven at the beginning of time, the sun, moon, and stars continue to give light to the earth, govern day and night, and separate the light from the darkness.[7] So reads the Old Testament.

The creation account in Genesis is revolutionary. We find in it no tale of battle among the gods; no supernatural warfare of the kind that appears in Babylonian, Greek, or Norse mythology. The innumerable gods found in the myths of every culture are reduced by the Jews to a single being: Jahve. God, in a solitary action spanning seven days, created heaven and earth, man and woman. While his original state was paradisaical, the acquisition of divine wisdom was, once again, joined with labor and loss. Eating the forbidden fruit of the tree of knowledge of good and evil meant for Adam and Eve the loss of Paradise and the beginning of want. To woman Jahve said, "in sorrow thou shalt bring forth children," and to man, "in the sweat of thy face shalt thou eat bread."[8]

In the Christian imagination, the Fall of Man was prompted by the temptation of Lucifer; humanity's ultimate redemption to come through Christ. In one apocryphal tradition, these two beings, Lucifer and Christ, were originally brothers, both high angels of light. A medieval maxim declared *Christus verus Lucifer*, Christ is the true Lucifer, which makes all the more sense as Lucifer's very name means "bearer of light." This being is no Ahriman, the dark spirit of denial from Zoroastrianism, but a being of radiant beauty who fell from the heights through the sin of pride. In Isaiah's words, he is "the bright son of the morning" who because he thought, "I will scale the heavens; I will set my throne high above the stars of God," was cast out of heaven. The story is told in the Jewish Books of Adam and Eve.[9]

God made man in his own image and blew the breath of life into him. Then came the archangel Michael and he brought Adam before the host of angels and bade them worship man in the sight of God as the Lord desired. Lucifer balked, responding: "I will not worship an inferior and younger being than I. I am his senior in Creation; before he was made was I already made. It is his duty to worship me."

For defying the warnings of Michael and the dictates of God, Lucifer was cast down onto the earth. There he saw Adam and Eve, who were still in Paradise. Miserable at his separation from God and envious of the happiness of the first pair, he beguiled Eve and through her Adam, so that they ate of the forbidden tree.

The Fall of Man was inseparable from the fall of the bearer of light, Lucifer. As angelic light was cast down onto the earth, man and woman ate the fruit of the tree of knowledge, and so were cast out to labor in the lands east of Eden.

The expulsion and fall of Lucifer, and the accompanying Fall of Man, finds its counterpoise in the incarnation of Christ—a luminous being of the sun. According to the gospel of John, Christ is "the light of the world," a light that shines into the darkness. John says it more bluntly still in his first letter. "This is the message we have heard from him and proclaim to you, that God is light and in him is no darkness at all . . . if we walk in the light, as he is the light, we have fellowship with one another."[10]

John the Baptist "came as a witness to testify to the light. . . . He was not himself the light; he came to bear witness to the light. The real light which enlightens every man was even then coming into the world."[11] Christ was to be understood as the "real light"—the true spiritual, consummately angelic light of the world. Without the burden of sin, that is, without expulsion from heaven, this high angel of light was nevertheless incarnated in the flesh of this earth. Resisting Lucifer's temptations to pride, he prepared for the shame and degradation of the Passion. Those who scorned him asked how could this be the Messiah, if the Lord God allowed him to be scourged, crowned with thorns, and crucified between two criminals? Yet in the Christian imagination, without these trials of sacrifice and love, performed in utmost humility, the light of resurrection could not shine. And might not the salvation of fallen man also entail the redemption of Lucifer—of fallen Light?

—

PROMETHEUS, TOO, WAS finally unbound from his agonies on the Caucasus by the hero Heracles. Having accomplished eleven of his twelve labors, which included an initiatory descent into hell and a rising again, Heracles journeyed to Mount Atlas in search of the golden apples of Hesperides. Near the end of that final task, he crossed the Caucasus and found and freed Prometheus from the agonies of his daily torture. Thus, the greatest hero of Greek myth, having completed his own soul's transformation through twelve mighty labors, could free the first benefactor of mankind. Himself free, Heracles could free others.

The shared, cultural imagination of man as sojourner in a dark, fallen world into which light has been mingled offered each soul the Heraclean task of world redemption, whether as a follower of Zoroaster or Christ. This imagination attained its most radical expression in the teaching of the third-century Iranian gnostic Mani.

Manichaeism—Religion of Light

Rich Friend of the beings of Light! In mercy grant me strength and succour me with every gift.[12]

Manichaean Hymn

In the main hall of a rustic, hillside shrine, thirty miles south of Ch'üan-chou on the eastern coast of China, is a stone statue of a Buddha-like figure.[13] Its various members have been skillfully carved from stones of differing hues so that the entire statue appears luminous. Many of its features are familiar to us from Buddhist statuary elsewhere; he sits in peace, cross-legged on a lotus, dressed in a *kasaya*, and enclosed in the sacred radiance of an embracing halo. Looking closer, however, we notice that this is no ordinary Buddha. In place of downcast eyes, curly hair, and the clean chin characteristic of a Buddha, we see long straight hair draping his shoulders, framing a bearded face. His penetrating eyes stare straight at the approaching pilgrim. Looking to the nearby inscription, we read: "Mani the Buddha of Light."

The origins of Manichaeism were far away from that unique Chinese shrine in time as well as space. When Mani was but twenty-four years old, "there flew down and appeared before me that most beautiful and greatest mirror-image of myself . . . my divine Twin."[14] From that day in mid-April A.D. 240 in the ancient city of Ctesiphon (near present-day Baghdad), there arose a religious movement that spanned the Roman empire, Persia, and the Far East. Mani, "the Envoy of Light," taught to his followers what his Twin had revealed to him, traveling widely, healing and teaching as he went. His was a religion of light whose central creation myth and ascetic practices acted so powerfully on the human imagination that Manichaeism spread without benefit of warfare, forced conversions, or active royal support. Perceived as a threat by every orthodoxy, Mani died a martyr's death, but not before he released what was the most widely held heretical vision ever to capture the imagination of man. Its last beautiful refrains were still being sung in the thirteenth century by the devoted followers of various Christian "Manichaean" heresies in southern Europe, until they, too, were ruthlessly destroyed.

It is not surprising that we find Manichaeism first arising in the countryside that fostered Zoroastrianism, for in many things they were similar. According to Mani, Light and Darkness were equal powers originally strictly separated from one another, but through repeated invasions they became mingled. When given the chance during the first conflict of the creation drama, the Prince of Darkness, Ahrmen, and his hosts seized upon and conquered the First Man, Ohrmizd, of the Realm of Light. By this means Light and Darkness were mixed and the world brought into existence. So, too, was the stage set for all future attempts at redemption. In order to liberate the First Man, the Father of Greatness evoked the Friend of Lights, who in turn issued the Living Spirit. Together with the Elements of Light, the Living Spirit entered the Realm of Darkness and liberated Ohrmizd, but in his flight Ohrmizd had to leave behind his luminous amour, the Five Elements. The purest particles of light from the Elements came to form the sun and moon, those slightly corrupted formed the stars, and a third of the Light was left behind, profoundly corrupted, to form the earth. In an attempt to thwart the redemptive efforts of the Beings of Light, the Prince of Darkness (not Jahve) created Adam and Eve in the image of the third divine messenger, the not-yet-incarnate Christ. (All this was taken as a fantasy heresy by the young orthodox Church.) Through Adam and Eve and their descendants, Light was to be trapped forever in the dark realm of Ahrmen. Mani's mission was its release. Light, imprisoned in the matrix of matter since the battles of Ahrmen, should make its way home again.

Within Manichaeism, the Elect (i.e., priests) had an especially important role to play in the project of redemption. For through their earthly conduct they could release Light and carry it within them through the gate of death. Even the food they ate contained enchanted Light. When rightly prepared by Hearers (those who heard the teaching but were not themselves Elect) and eaten by the Elect, the Light within it would be released into the priests' care. At their death the Elect journeyed in spirit to the Moon with their harvest of Light. The Moon waxed with the bounty brought by them until, when full, a Column of Light would rise from the Moon and streamed on for a final return to the Sun, its proper celestial home.

During his life, Mani had traveled widely, preaching his message of Light and Darkness, of knowledge (*gnosis*) and salvation. He envisioned a universal religion that brought together the heritages of Buddha, Jesus, and Zoroaster. Manichaeism's influence was enormous. One of the greatest figures of early Christianity, St. Augustine himself, was a Manichaean from 373 to 382, and although later he vigorously attacked its doctrines, his thinking was never really free of its effects, and indirectly through his writings a metaphysics of light, born of Mani's vision, made its way into orthodox Christian thinking.

As Christ, Light incarnate, was delivered by jealous Jewish priests to Pilate for execution, so was Mani, the Envoy or Buddha of Light, given to the Persian king Vahram I for execution at the instigation of the zealous Zoroastrian priest Kirdir, who tolerated no rivals. Mani, however, was confident that his teachings would live on long after his death (in A.D. 277), and so they did.

Throughout the history of the early Church, Mani's cosmogony of light, and tenets of many kindred heresies, constantly challenged orthodox doctrines. Their elaborate imagination of our universe as caught in the struggle between spiritual light and ponderous darkness informed the lives of countless followers. In addition, the contemporary Dutch theologian G. Van Groningen sees a less personal but equally effective influence of gnostic thinking. In the gnosticism of Mani and similar sects, Van Groningen detects an earnest concern with the scientific and philosophical issues that take their point of departure from man.[15] These sects, not the early Church, were deeply committed to knowledge. They aspired to a knowledge or *gnosis* of the spiritual cosmos not in order to master the earth, but to free the soul that it might return to its native realm of light. Still, gnostic aspirations, Van Groningen suggests, foreshadow scientific ones. The early Church saw in such striving the luciferic sin of pride. Salvation, after all, was given by grace alone, not by knowledge. Through the Middle Ages the western Church carefully restrained those in its fold whose heterodox studies (and there certainly were such) might take them too far. Others outside the fold were treated more harshly. Yet, as he had predicted, Mani's impulse lived on. The last dramatic scene in that history was the flowering and destruction of the Cathars, or "Pure Ones," of thirteenth-century southern France.

Between the villages, châteaus, cities, castles, and throughout the countryside of Languedoc, lean, pious men journeyed, two by two, long-haired and dressed in black. Gaunt from hard fasting, they spoke quietly, without anger or worldliness, of Christ and a path of light, the path of salvation. Many were their acts of charity, and constant was their attention to God and a pure life. To the young and simple they told fairy tales of extraordinary beauty;[16] to others they gave religious instruction according to their place within the community. In devout simplicity, they celebrated together the rituals and sacraments of their church. To the people they were known as *bons hommes*, or "good men"; within their religion they were *perfecti*. Passing an elderly *perfectus* who was ill-clad and crippled, the Count of Toulouse said: "I would rather be this man than a king or emperor."[17]

While, in fact, distinct from Manichaeism, the Cathars (and related heresies of the period) shared the Manichaean view that the world was the creation of Satan, and that through their righteous conduct, the divine Light within them could be saved. Failing that, they would return to life in the form of another sentient being (metempsychosis) to try again.

The established Church of Rome first ignored and then fought, with more and greater force, the incursions of the Cathars into its dominion. The Church sent representatives from its own newly established mendicant orders, most notably St. Bernard in 1145 and St. Dominic in 1205. Trudging barefoot under the hot sun and dust, St. Dominic begged for alms and an audience. Tormented wherever he went, ridiculed and mocked, he preached the doctrines of the Catholic Church not only in his words, but also in his actions of devotion and poverty. While these certainly made an impression on his hearers, neither he nor St. Bernard had much success in converting Cathars. When they failed, repression was accomplished through violence. While St. Francis, in Assisi, was writing his beautiful *Canticle to the Sun,* a glorious paean to all existence, the Cathars were being ruthlessly destroyed by sword and fire in southern France.[18]

The final blow came in 1238 with the fall of the Cathar High Place, the mountaintop fortress at Montsegur. Two hundred *perfecti* sheltered there were burned without trial. While it took another century of systematic repression, execution, and brutality to extinguish the ideal of "purity" espoused by the Cathars, by 1330 the Cathar church in France was no more. St. Bernard of Clairvaux, who fought the Cathars with all his strength, said of them, "No sermons are more thoroughly Christian than theirs, and their morals are pure."[19] Their austere heresy took with consummate seriousness the reality of angelic light in a darkened world. They were the last in the West to live fully within an imagination of spiritual light. With the eradication of the Manichaean heresy, a mighty imaginative stream ended, effectively disappearing from the stage of external history.

Yet the metaphysics of light espoused by Manichaeism found

a more orthodox echo in the religious and scientific convictions of the thirteenth century's most significant figure in the biography of light.

Robert Grosseteste's Cosmogony of Light

In 1229, just three years after the death of their beloved St. Francis of Assisi, the English Franciscans invited the elderly churchman and scholar the Archdeacon of Leicester, Robert Grosseteste, to teach the young friars at their newly built Oxford school.[20] Six years later, in 1235, he was consecrated Bishop of Lincoln, the largest diocese in England. As teacher and then as bishop, Grosseteste was famous for his upright and moral life in an age when clergy were often more interested in temporal delights than the final rewards of the hereafter. Grosseteste had a special regard for the poor Franciscans who shunned the riches of the Church for those of the soul, and they returned his affections in kind. Combined with his upright character was a love of music, a sharp wit, a devotion to learning, and a single-minded preoccupation with the nature of light.

Among Grosseteste's books, one is a gemstone that outshines all the rest. It is the slender volume entitled *De Luce*, or "On Light." In order to appreciate it we need to remind ourselves of Plato's influence on medieval scientific thought.

Near the end of his life, Plato wrote a lengthy dialogue *Timaeus*, which begins by recounting a conversation between an aged Egyptian priest and the Greek poet Solon in the holy city of Sais. The priest, having access to documents of great antiquity, described to Solon the most ancient history of Greece, going back long before the deluge. Timaeus, the astronomer in Socrates' group, continued the tale by going back even further, giving an account of the generation of the universe. The "likely story" told by Timaeus is of the creation of the world by God through his subordinate Demiurge. Yet its character is less a religious than a scientific presentation, relying on reason, evidence, and speculation, and not on divine inspiration. The world is formed according to number, ratio, and geometry, wrote Plato. The five elements of earth, water, air, fire, and ether, or quintessence, are composed of triangles that join to form beautiful solids: the five regular "Platonic solids." His cosmology was thoroughly mathematical, and in Greek times that meant primarily geometrical. Throughout the Middle Ages the *Timaeus*, perhaps more than any other document, shaped the scientific imagination of the West, and together with Genesis challenged medieval thinkers to fashion their own "likely story" of the world's beginnings. The *Timaeus* was one of the few texts of Greek thought that had made its way, at least in partial translation, into the early intellectual life of Western Europe.

Grosseteste's *De Luce* is the only comparable work of scientific cosmogony between the time of Plato's *Timaeus* and the eighteenth century. It is a watershed document. On the one side, receding into a haze of unclarity, rise the magnificent heights of prescientific Greek and Christian thought; on the other are the as-yet-unborn achievements of modern science and technology. Some years ago the eminent French historian Alexander Koyré advanced the striking opinion that *De Luce* was the first step toward the establishment of a modern mathematical science of nature.[21] On opening the book, we therefore would expect to see pages dense with mathematics, calculations, and geometrical figures. Instead, we find a story of creation, a Neoplatonic genesis in which the entire universe comes into being through the expansion and modification of light.

Light, according to Grosseteste, was the first form of corporeity, and from it all else followed. Multiplying itself from a single point infinitely and equally on all sides, light formed a sphere and together with this action arose matter. Following the expansion of light to its ultimate extent came a phase of differentiation caused by condensation and rarefaction, which led in turn to the separation of heavens and earth and to the origin of the thirteen spheres, nine heavenly and four mundane. We will not follow Grosseteste's arguments in detail, but I would emphasize the twin themes of light and measure (or geometry) that pervade the work.[22] Like Plato, he embedded within his cosmogony his vision of God as Geometer and Numerator, who "ordered all things by number, weight, and measure."[23] The medium chosen by God for his creation was light. Yes, *De Luce* offers us the tender beginnings of mathematical science, but everything is implicit, contained in its structure, style, and Grosseteste's manner of inquiry.

All of material creation, then, is condensed light. The modern architect Louis Kahn was unwittingly paraphrasing Grosseteste when he said to an interviewer for *Time* magazine, "We are actually born out of light, you might say. I believe light is the maker of all material. Material is spent light."[24] Yet according to Grosseteste, the condensation of light is only one pole of God's creative action, the pole leading to sensual creation. Grosseteste also believed deeply in the supersensible pole of creation, in angelic light. According to his view, we stand upon the earth surrounded by the four visible kingdoms of nature (mineral, plant, animal, and man), but above man rise the nine hierarchies of angelic beings: Angels, Archangels, Archaei, Exusiai, Kyriotetes, Dynamis, Thrones, Cherubim, and Seraphim.

Together with the recognizably scientific aspects of his cosmogony of light, Grosseteste embraced an explicitly spiritual metaphysics of light. To Grosseteste, God's declaration "Let there be light" possessed two aspects. One aspect would ultimately become the light of our physical existence condensing even to the form of matter, but the other aspect was a light of intelligence embodied in the purely spiritual, angelic creations of God. In his commentary on the opening chapters of Genesis, and especially in his detailed studies of the four major writings of Pseudo-Dionysus, Grosseteste revealed his profound interest in the spiritual ordering of our cosmos, and once again he saw it as best symbolized as a cosmos of light, but now angelic light.

In Grosseteste, we also find the first advocate for mathematics as a pivotal element of the new science. He wrote, "nothing magnificent in [the sciences] can be known without mathematics."[25] Grosseteste has been called the first expositor of experimental science. He provided, at least in theory, the twin pillars essential to the development of modern science: mathematics and experimentation. A. C. Crombie declared him "the first to set out a systematic and coherent theory of experimental investigation and rational explanation by which the Greek geometrical method was turned into modern experimental science."[26] Others followed who would actually enact the program Grosseteste described, but he articulated with forceful clarity the tenets that would become basic to the new science over the following centuries. Both the metaphysics and physics of light played through his mind with equal ease and honesty. It was, after all, God's created universe, material and spiritual, and the place of man as a citizen within both called for knowledge of both realms whether through revelation, reason, or experimentation.

Grosseteste's thinking had been influenced not only by Platonism and Christian theology, but also by the first great philosopher of the Islamic world, al-Kindi. During the ninth century, al-Kindi had not only vigorously argued for an improved version of Euclid's extramission theory of sight, but also sought to integrate it with his spiritual philosophy of nature. The eye is not alone in sending forth rays, but "everything in this world produces rays in its own manner like a star. . . . Everything that has actual existence in the world of the elements emits rays in every direction, which fill the whole world."[27] The universe of al-Kindi was woven together by a vast web of lightlike rays linking stars to earth, magnets, fire, sound, and so on. Optics thus became an archetypal image of a universal process: the radiation of power. Echoing the cosmology of al-Kindi, Grosseteste saw the sun and moon in the eyes of the human face; "for the head is borne towards the heavens and has two lights, as it were sun and moon."[28]

Robert Grosseteste is for us a figure who stands, Janus-headed, facing in two directions. One face gazes upward and back, evoking a metaphysics of light that embraces hosts of angelic beings and emanations of light that are active in God's creation as intermediaries and enactors of His will. His other face gazes earthward to a future when the physics of light will develop to its fullest extent as modern optics, and most especially as a mathematical physics grounded in experimental observation. Both faces, both of Grosseteste's enthusiasms, are rightfully a part of our study of light.

—

LIGHT WAS ONCE the sight of God. As the gaze of Ra spanned space light stretched from one corner of the universe to the other. In the night sky, planets and stars once played host to gods and angels, who in turn passed on their gift of light to man through mishap or cunning. In Christianity, the greatest cosmic being of light, Christ, ancient brother to Lucifer, came down to earth. It was as if the Persian sun god Ahura Mazda were to don vestments of clay. The cultic imagination of light reached its highest form in the gnostic-Manichaean sects in which the redemption of the world was the release of light through human actions. The final grand inflorescence of this religious imagination perished in the fires that consumed the Cathar *perfecti*.

Henceforth light would grow to be understood scientifically. Arising from Greek philosophy, moving through Persian and Arab lands to the desk of Grosseteste and his contemporaries, man's imagination of light profoundly changed, becoming ultimately the conceptual basis for the optical sciences of modern physics. Grosseteste still felt the presence of a glorious spiritual tradition that could nourish his inner life. He sought to mate the first buds of natural science with his religious thought in a metaphysics of light. With the passing of the Cathars and of Grosseteste, the religious tradition of angelic light faded. Over time, science pruned away the trappings of spirit to fashion a material and mathematical imagination of light. In doing so, it similarly reshaped its image of man and cosmos.

CHAPTER 4

The Anatomy of Light

Hail holy Light, offspring of Heaven first-
  born!
Or of th' Eternal coeternal beam
May I express thee unblamed? Since God is
  light,
And never but in unapproached light
Dwelt from eternity, dwelt then in thee,
Bright effluence of bright essence increate!

Milton, Paradise Lost

On the Threshold of Scientific Sight

Outside the sublime cathedral of Santa Maria del Fiore in Florence, the day was lovely and a light wind pushed billowing white clouds across an azure sky. It was exactly the kind of day Filippo Brunelleschi had been waiting for, so he let it be known among his friends that a long-promised demonstration would take place in the doorway of the cathedral. [1]

Standing on the threshold, Brunelleschi looked across the cobbled piazza to the beautiful Baptistery of San Giovanni, already three hundred years old. Within its dim interior, glorious mosaics hovered overhead on its domed vault, including the huge figure of Christ at the Last Judgment. Dante Alighieri had been one of the innumerable Florentines brought into the community of the faithful through baptism there. Just now, however, standing on the threshold of the cathedral, Brunelleschi did not see the sacred history of the baptistery, but rather a network of lines receding to an infinite horizon. He looked on the piazza with the eyes of the great inquirers who would come later: Galileo, Descartes, and Newton. For today he was not only architect and sculptor, but also geometer and optical scientist.

His friends gathered around him, the slight youthful artisan, and listened intently to his animated explanation of the demonstration he would perform before them. A genius, he would go on to design a great dome for the cathedral behind him—an enormous task that would last sixteen years, employing scores of craftsmen and artists and consuming vast sums of money. His task today, however, was far more modest in budget and scale. It required only a wooden panel twelve inches square, a mirror, the brushes of a miniaturist, paints, and a single man. Yet in that afternoon, Brunelleschi performed magic, a trick whose consequences would reach further by far than those of his dome, or indeed anything else he would ever create. Quite simply, he was changing the way we see.

On the threshold of the cathedral, Brunelleschi made the first drawing in linear perspective. Scholars place the date between 1412 and 1425.[2] For the first time in history, man was creating a scene so realistic that when it was viewed from the right place, "the spectator felt he saw the actual scene" before him, as Brunelleschi's contemporary Manetti tells us.[3]

In order to prove the power of perspective illusion, Brunelleschi contrived an ingenious demonstration. Those present that day viewed his perspective rendering of the Florentine baptistery in a bizarre but highly dramatic way. (See the figure on page 60.)

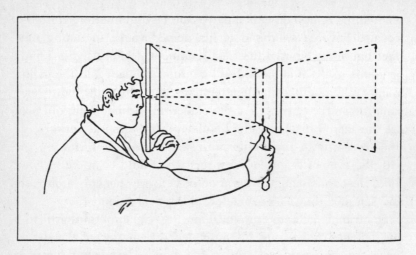

Brunelleschi's illusion of perfect perspective. A mirror in his right hand and the painting in his left, Brunelleschi looked through a peephole toward the baptistery. With or without the mirror, the scene was the same.

They stood in the door of the cathedral, facing the baptistery. We should imagine Brunelleschi's friends holding the painting, one by one, with its painted surface away from them, and peeping through a hole drilled through its center. Through the hole they saw the piazza and baptistery. Into the line of view they now inserted a mirror whose size and shape were such that, when held at arm's length, the reflected image of the painting just filled the mirror. But the painted scene viewed in reflection was *exactly* that seen directly. In other words, the viewer saw the same thing with and without the mirror: a perfect illusion. To enhance the illusion further, Brunelleschi made the painting's sky out of burnished silver, so highly polished that its mirror surface reflected the windswept clouds as they drifted overhead. The illusion was masterful and complete.

The effect was stunning, and its consequences were destined to dominate all of painting until the twentieth century. How had he done it? Why hadn't it been done before? Euclid's theory of vision was completely adequate to the task and had been known for centuries. Yet something had been lacking, not some new fact but a new spirit, a new scientific way of seeing.

A Spiritual Geometry

For the mind, only that can be visible which has some definite form; but every form of existence has its source in some peculiar way of seeing, some intellectual formulation and intuition of meaning.[4]

Ernst Cassirer

In a small room of the Frick mansion in New York hangs a painted panel not much larger than the one by Brunelleschi in the doorway of the Duomo. It is a work by Duccio painted in 1310 depicting the temptation of Christ by Satan. Two enormous figures, Christ rejecting his dark adversary, stand on a stylized mountain amid a landscape barren but for a few surrounding clusters of buildings. From the standpoint of linear perspective or the way a camera would have caught the scene, nothing is correct in the painting; but from the standpoint of spiritual relationships, the arrangement is perfect. The sky, made of gold leaf, suggests that we are not looking into a physical but rather a moral or spiritual space. Everything—the colors, the gestures, the relative scale and positions of each figure, building, and mountain—reveals a sacred order that may well be at variance with the laws of vision, but is in true harmony with the values of fourteenth-century Italy. In just this way did the painter, and the painting's viewers in Siena, see their world.

What mattered most for them was not the physical but rather the spiritual geometry of space. This had been true of Egyptian art and it remained true of Duccio's. Pharaoh and the Madonna had to be larger than those who served them because their spiritual stature required it. Painting depicted reality, and in

The Temptation of Christ on the Mountain by Duccio.

doing so it created a language of expression in profound accord with human experience, not the laws of optics. Only when "what was" changed, when Brunelleschi and his associates began to experience their world from the standpoint of Euclidian analysis, did the means arise to represent their vision: linear perspective.

It is difficult, almost inconceivable, for us to imagine a culture that does not see pictorial images as we do. Perhaps an example from cross-cultural studies will help. Twenty years ago, two acquaintances of mine worked in remote villages in Africa with indigenous peoples. Their project was to improve infant care in the hopes of lowering the staggering infant mortality rate. One friend was a professional photographer and so developed a slide show on infant care. Once it was perfected to his satisfaction, he packed his high-tech presentation into his truck and headed out to the bush to join the woman who is now his wife. After some efforts they devised a way to show the slides to women of the village. At the end he asked them how they liked the presentation. Very nice, the women said. The colors were beautiful, and the shapes were very interesting. After a few conversations, it dawned on him that during the evening's performance much or all of the slide imagery was simply *not seen* as he had seen it. The colors were great, but they didn't see the images as representative of infants, objects, or people. The context and scale of the images were so remote from their experience as to make them nearly meaningless. This single instance could be multiplied endlessly with startling and confusing anecdotes from early missionaries, anthropologists, and explorers.

In a more systematic cross-cultural study of pictorial perception, Jan Deregowski studied the perceptual abilities of Zambian children and adults.[5] One aspect he studied was the assignment of spatial arrangements and depth to pictorial images. From his work and others, it is clear that many visual clues are ignored or misinterpreted (from our viewpoint) in other cultures. Consider the phenomenon of "split type" drawing. When African children and adults were presented with a perspective drawing of an elephant from above, and another showing the elephant unnaturally split, also seen from above, they almost universally preferred the split-type drawing.

The "properly" drawn perspective view of the elephant was, according to them, "jumping about dangerously." The preference for split-type drawing is remarkably widespread, according to anthropological studies. When representing an animal or face, the African artist, like the child or pre-Renaissance artist, could

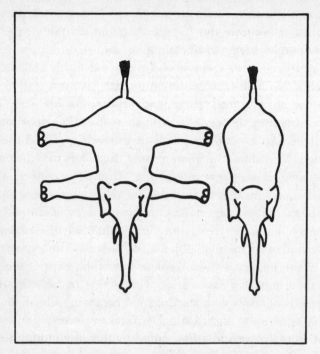

The "split" elephant was preferred by African children
and adults to the top-view, perspective drawing.

not leave out the "hidden" side. Our sense that linear perspective renders a scene truly is purely cultural. It can only appear when we, as a culture, have sufficiently deadened the active light of interpretation. Then the eye or legs that are hidden but that we know are there can be left out; the relative size of pharaoh can be the same as or even smaller than that of his servant, even though we know he is the greater being. One can almost feel the withdrawal of interpretive light with the birth of linear perspective. Yet even here, our mind is still active, otherwise we would struggle, as Moreau's child did, to see the world.

From these and our previous considerations of sight, we realize that every age and culture has crafted its own sensory reality, forged its own view. We may apply the same lesson to the discovery of linear perspective.[6]

The pre-Renaissance history of perspective is complex and controversial. As the eminent art historian Erwin Panofsky has argued, "each period in western civilization had its own 'perspective,' a particular symbolic form (as Ernst Cassirer called them) reflecting a particular *Weltanschauung* or world view."[7] Elements of perspective can be found dating back to the Greeks, but not until 1425 did a coherent, clear understanding of linear perspective appear. Brunelleschi and the Renaissance longed to see and represent the world in accord with "visual truth," as Lorenzo Ghiberti called it.[8] They "endeavored to imitate nature," to school their sight on the mathematical lawfulness of nature. They incorporated into themselves the geometric spirit that breathed through the age and thereby helped lay the foundation for the style of scientific investigation to follow.

By 1525 the spirit of artistic-scientific perception had so developed that Albrecht Dürer could publish his unique *Painter's Manual*. Opening it, one does not find a single section on color or painting as such. Rather, it begins with instructions concerning the geometrical construction of spirals and conic sections, then goes on to polygons, the Pythagorean theorem, towers, sundials, and regular and irregular solids. There is even a detailed section on the "Construction of the Alphabet" from arcs, circles, and lines. Nowhere is the spirit of the manual, and so of the period, more evident than in the final section on linear perspective.

The principles of linear perspective derive from Euclid's concepts of the visual ray and the visual cone as elaborated by Ptolemy and al-Kindi. Rays from the eye extend to every point of the seen object, forming thereby an irregular cone with its apex at the eye. Imagine a plane surface (the painter's canvas for example) somewhere between the object and the eye. Each

Goddess and craftsman. A perspective drawing machine by Albrecht Dürer.

ray will pass through the canvas at a particular point. Suitable markings of each point on the canvas will create an accurate representation of the object in perfect perspective.

In his *Painter's Manual,* Dürer includes descriptions and drawings of various devices that permit the draftsman to apply Euclid's theory and so to make completely accurate perspective drawings. The most revealing device is his last, wonderfully illustrated by Dürer himself. On the left of the drawing is a woman, unclothed but for a carefully placed drapery, reclining as a goddess might on a sea of pillows. Perhaps it is a study of Aphrodite, the beautiful and wanton goddess of love who, curiously, was wife to fiery Hephaestus, the lame god of craft. Yet something is amiss. She lies not on a couch but on an optical bench outfitted for "projection and section," as it would be called today. On the right is the draftsman whose eye is positioned steadily behind a pointer. From it his line of sight (Euclid's cone of visual rays) is directed to the object (that is, the "goddess") through a frame crisscrossed with stout black thread. The grid so formed acts as a coordinate system superimposed on the sensuous figure of the reclining nude. Point by point, the draftsman, with utmost visual acuity, transfers her divine figure to the cross-ruled paper on the table before him. Recall that young Tiresias had been blinded by a sight such as this. To the right, however, we have no youth but a descendant of Aphrodite's husband, Hephaestus the divine craftsman.

The juxtaposition is remarkable. To the left is a figure whose pose evokes erotic emotions or sentiments of sublime sensuality that had long been part of the artistic tradition. To the right is the dispassionate, objective eye of the draftsman recording in precise detail every feature of the body before him. It is the meeting of two streams of time, one flowing from out of the past, a stream of beauty both sensual and spiritual, and a second stream flowing from out of the future, one that aspires to clarity and control. They are wed in the picture plane of the artist's canvas as Aphrodite was wed to Hephaestus.

Nor is Albrecht Dürer alone in his enthusiasms. The most powerful image I know that reflects the shift in consciousness then under way is Leonardo da Vinci's anatomical study of a man and woman in the act of love, of procreation, dating from the first years of the sixteenth century.

Human dissection had long been under religious and popular prohibitions, only gaining legality in the century before Leonardo. The story of Michelangelo's secretive anatomical studies is well known. Under cover of night he would steal into the morgue in order to dissect the unclaimed bodies of beggars by the light of a flickering candle. Until Leonardo's time, little new, attentive work had been done in human anatomy. The medical texts of antiquity were given prominence over original anatomical study and illustration. Leonardo was certainly the most brilliant anatomist-artist of the period.

In his anatomical study of human sexual intercourse, Leonardo takes the consummate act of intimate human affection and displays it in anatomical cross section; the bodies of the two lovers are cleaved from head to groin. The inquisitive eye of the artist-anatomist has fixed on this action as it might on any other and submitted it to the clear, calm gaze of the new spirit. The drawings are still suffused with an artistic tenderness not to be found in modern anatomical texts. We should not

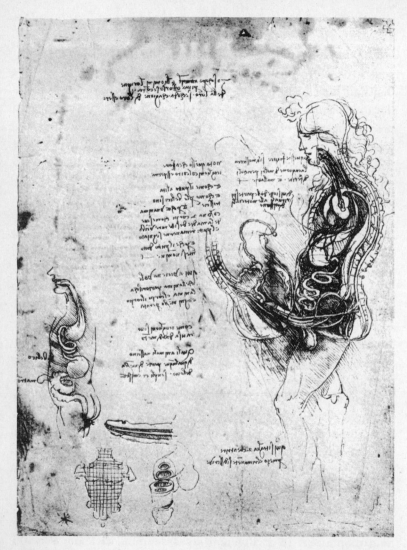

The anatomy of love, from the anatomical drawings of Leonardo da Vinci.

imagine Leonardo, Dürer, and Brunelleschi as revolutionaries bent on the overthrow of a godly universe. They viewed their efforts as a service to God who Himself had created the world according to number and measure. Leonardo longed to see deeper than the surface appearance of things, and hoped "that it might please our Creator that I were able to reveal the nature of man and his customs even as I describe his figure."[9] The Gothic cathedral had also been the expression of a conviction that God in his creation worked geometrically.[10] Renaissance painters saw perspective painting as a similar imitation of His work.

—

THE PHILOSOPHICAL ORIGINS of the new attitude to nature can be found in the remarkable thought of cardinal and mathematician Nicolaus of Cusa in the fifteenth century.[11] For hundreds of years nature had been alternately condemned as the product of the Fall, man's expulsion from the Garden of Eden, and mystically revered as the work of an almighty Creator. By the thirteenth century, the tide shifted toward a view of the earth as a glorious illustration of God's wisdom and power. The Lord had given man two books, it was said: the Bible, which was written by the "light of revelation," and Nature, written by a second light, the "light of experience." Both were expressions of the same hand. The study of scripture and the philosophical investigation of nature might be understood as two roads leading back to the same ultimate source.

Nicolaus of Cusa took matters a significant step further by likening God to the then-novel mathematical concept of infinity. Man is a finite creature, and since the multiplication of a finite number by any other is always finite as well, God (as infinity) is forever unattainable. The ratio of any finite number to infinity is always infinite, no matter how great the finite number. Up to this time the cosmos of the medieval mind had been a continuous chain of beings (Angels, Archangels . . .) stretching from man upon the earth to the heights of heaven and God. Grosseteste had certainly still embraced such a vision, but with Nicolaus of Cusa the chain was broken. As if in a second Fall, man and woman were separated further still from the divine. Cusa observed that man, finite being that he is, can never attain to the Godhead, no matter how many rungs he climbs on the ladder leading to the spirit. Mathematical logic had proven that our slightest imperfection places us infinitely far from God. Better then to study the earth, the finite creation of God, than to presume to moral perfectibility and proximity to the Godhead. For centuries, humankind, eyes fixed on the sun and stars, had aspired to become gods. Then, quite suddenly, the last hope was taken away for a direct ascent, and we had to be content to study the reflected luster of God in earthly knowledge.

The light of nature, therefore, became an ever-more-trustworthy source of human understanding, and Renaissance artists such as Leonardo da Vinci, Brunelleschi, and Albrecht Dürer stood ready to see by its light. Bookish scholarship was spurned: "If indeed I have no power to quote from authors," wrote Leonardo, "it is a far bigger and more worthy thing to read by the light of experience which is the instructress of their masters."[12] Moreover, following Nicolaus of Cusa, mathematics brought with it the surety now felt to be lacking in theological or philological analysis, and once again Leonardo was in the lead: "There is no certainty where one can not apply any of the mathematical sciences."[13]

The enormity of this transition should not be missed. We witness here other faces of the paradigm shift described by Thomas Kuhn in his *Structure of Scientific Revolution*. As Albrecht Dürer looked through the Cartesian latticework of threads at a reclining Aphrodite, and as Leonardo da Vinci sketched the act of love in anatomical cross section, their way of seeing passed from that characterizing one era to that of another. They crossed over the threshold to scientific sight. In the two hundred years from Duccio to Dürer, Western man moved from a religious perspective to a scientific one, from a moral to a material universe. Brunelleschi had strode from the sacred, interior space of the Duomo in Florence to its doorway, to the threshold of a new vision. His successors passed from the doorway into the lively, secular space of the piazza. Initially there lingered the memory, the sounds and smells of the rituals performed inside, but the life of the piazza held its own fascination. Its clamorous voices did not sing psalms in church Latin, but called out in the Italian used by Leonardo and Galileo. Its offering was not the body of Christ, but wares suited to worldly needs. What seemed beyond reach in one domain could be acquired in another, if one but had the currency of the marketplace.

—

CONTINUING ACROSS THE piazza of Santa Maria del Fiore, one comes to the east door of the baptistery, the "Door of Paradise" as Michelangelo called it. Commissioned of Lorenzo Ghiberti in 1425, about the time of Brunelleschi's demonstration, it includes the exquisite panel of the wise King Solomon receiving the beautiful Queen of Sheba in his resplendent temple, boldly executed in the new linear perspective. Solomon's temple had always been the biblical archetype for church architecture, the inspiration of architect and mason. The temple's master builder, Hiram of the line of Tubal-Cain, is not shown in the company of the priestly king and queen. He remains, like so many other master craftsmen, unseen.[14] Yet the times are changing. The marriage taking place in fifteenth-century Florence and elsewhere is not between secular and religious powers, but between *techne* and *sophia*, between craft and wisdom.

Until now, learning required leisure, and leisure required slaves or the support of the Church. Scholars and priests spurned the manual arts as beneath the dignity of their noble aspirations. Yet the mood of the time was shifting. After all, God, too, had

The marriage of the Queen of Sheba to Solomon the Wise, rendered with exquisite linear perspective in bronze by Ghiberti.

labored for six days in creating this world, and Jesus was a carpenter. Human labor might be more than penance, it might even be the imitation of God.[15] Monastic life changed during the Middle Ages, coming to value work as prayer, and thereby helping to form a climate of thought where Hiram could appear at the flank of the Queen of Sheba. I imagine her as an intermediary between the priestly wisdom of Solomon and the earthly forces, of which Hiram was supreme master.

From the union of scholar and craftsman, there flowed a new stream. Its current carved the canyons we now inhabit. Galileo was one of the early merchants to ply its waters, and one of the things he sought was the body of light.

The Perfect Argument

SAGREDO: I cannot without great astonishment hear it attributed as a prime perfection and nobility of the bodies of the universe that they are invariant, immutable, and inalterable. For my part I consider the earth very noble and admirable precisely because of the diverse alterations, changes, generations, etc., that occur in it incessantly.[16]

Galileo, 1632

In his three monumental compositions, *Paradiso, Purgatorio,* and *Inferno*, Dante had given detailed information on the disposition of heaven, the anguish of souls in purgatory, and the eternal damnation of the wicked. In view of Galileo Galilei's ultimate condemnation by the Inquisition, it is curiously ironic that in 1588 the Florentine Academy invited him, then a gifted young scholar and mathematician of but twenty-four years, to address its august body on the location, size, and arrangement of hell as described in the *Inferno*.[17] In retrospect it seems an unlikely subject for the young man who was later to lecture the world on other topics, such as the mathematics of projectile motion, the construction of the telescope, and his revolutionary observations made with it. Perhaps the later condemnation of an aged and blind Galileo by the Holy See was intended to speed him on his way to eternal rehabilitation in a place whose dimensions and accommodations would be well known to him from his youthful research. His own position was that "the holy Bible and the phenomena of nature proceed alike from the divine Word, the former as the dictate of the Holy Ghost and the latter as the observant executrix of God's commands."[18] Thus no contradiction could exist between the two books (scripture and nature) if both were read correctly, and no damnable violation of God's trust was performed in attending to and experimenting with the book of nature.

What then was the argument between Galileo and his implacable opponents? In large part it was over the location of perfection. Aristotle in his *On the Heavens* had stated the matter clearly. The earth, immobile and at the center of the universe, was surrounded onionlike by sheaths first of the four elements (earth, water, air, and fire, in that order), and then by the seven planets: moon, Venus, Mercury, sun, Mars, Jupiter, and Saturn. From the moon out, the universe was made of an incorruptible "quintessence" or fifth substance, whose sole movement was perfect circularity. All heavenly objects, quintessential in nature, should also, therefore, have a perfect and unblemished shape—the sphere. Moon, sun, planets, and stars were conceived of and "seen" as polished, transcendent, spherical objects unsullied by the dirt, chasms, and mountains of the "sublunar" realm we inhabit. Perfection was *up*, rank upon rank above man, as Dante, Grosseteste, Dionysius, and myriad others had imagined, in a great angelic chain of being, now stiffly reflected in the incorruptible, crystalline spheres of the planets.

It seemed axiomatic that spiritual perfection was reflected in physical perfection. Odysseus, washed up naked and briny on the Phaiakian shore of the island of Skeria, "looks like one of heaven's people" after a quick bath.[19] His uncertain origins were recognized as noble by his skill in games, casting his discus many times farther than all the Phaiakians.[20] The spiritual perfection of sun, moon, and stars should, likewise, find full reflection in their physical attributes: unblemished, spherical, and eternal.

In 1609 Galileo heard that "perspective glasses," what we would call telescopes, had been made by the Dutch spectacle makers of Middelburg. Galileo investigated the properties of various lens combinations and constructed several improved telescopes for himself. Pointing his novel device at the heavens, he invaded sacred space with a secular and scientific mind. Unlike his contemporaries, Galileo was not content with leaving the sanctuary of the Church for the bustle of the piazza, but insisted on bringing the money changers into the temple. Gazing "heavenward," he saw, instead of angels and perfection, craters and mountains on what had once been the mirror surface of the moon. New planets (moons) circled distant Jupiter, and even the sun, that source of God's pure light, was marred by hideous blemishes—sunspots. Perfection lost. No longer could one look up to it on a starry night, no longer did the gods or God illumine the universe as Helios. The sun was indeed a "fiery rock," as Anaxagoras had prophetically declared two thousand years earlier. The analytic and earthen eye—Brunelleschi's, Leonardo's, Dürer's, and Galileo's—had reached out and touched the heavenly spheres. The skies' crystalline purity turned to common stone, and the ground shook under both the university and the Church, always the great bastions of conservative thought. In the light of Galileo's discoveries, where could one now hope to find perfection?

Galileo could answer that he had not destroyed perfection, but only displaced it to its rightful location. In exploding the myth of incorruptible heavens, he moved perfection to the mind and the insubstantial universe of pure mathematics. The earth and its manifold creatures had always been seen as transient and corrupt; now the entire universe shared the fate of man—imperfection. Pure thought alone, mathematics, remained inviolate. By unbelievable coincidence, this imperfect world had hidden beneath its scaly skin the pristine forms of mathematics as laws of nature. As Galileo's contemporary Johannes Kepler declared, "Geometrical reasons are co-eternal with God."[21] From the time of the ancient Pythagoreans, and before them the astronomer-priests of Babylon, men had seen number in the world. Freed from religious constraints, that mathematical view erupts in the sixteenth century to become one of the two main currents in scientific thought.

While perfection was becoming disembodied a second and equally essential stream developed in Galileo's mind as if to balance the first. No longer was Aristotelian physics tolerable with its four elements, doctrines of matter, form, potentiality, and actuality, etc. In its place should stand clear causal accounts, understood to be material and mechanical in character, for every phenomenon.

Views such as these had been advanced in antiquity by Greek atomists, but they were abjured as atheistical and dangerous. Plato in the *Laws* of his ideal city reserved a specific punishment for those who held and promulgated such views.[22] They were to be taken to the desert, placed in solitary confinement on bread and water, and taught the error of their ways. If unrepentant, they were to be executed. All this from a philosopher whose own beloved teacher, Socrates, had been sentenced to death for corrupting the youth of Athens. Galileo was not sentenced to death for displacing perfection; he had powerful supporters, and was politic enough to dissemble to the Inquisition while laboring furtively to publish his heretical views.

If Aristotle was to be set aside, what could act as the basis for secure knowledge? If one wished to begin with sense impressions, how could one move beyond their often misleading report? In earlier times, one saw through them the indirect messages of the gods. What stood behind them now? Especially in his early writings, Galileo's position was that sense impressions were purely subjective, that "tastes, odors, colors, and so on were no more than mere names so far as the object in which we place them is concerned, and they reside only in consciousness."[23] For example, something we experience as hot is really only "a multitude of minute particles having certain shapes and moving with certain velocities." Such is heat. Although Galileo appeared cautious in his speculations about an ultimate materialistic, corpuscular basis for the world, later philosophers and scientists felt no such scruples. If a divine world existed, it was severely separate from the material world of the senses, and little or nothing stood between them.

—

PLATO HAD ARGUED in his cosmological dialogue the *Timaeus* that between two numbers a middle term could always be found that mediated and made more beautiful the mathematical relationship they expressed. His Demiurge created the world according to such a threefold relationship. In the graduated cosmos, intermediaries had connected earth and heaven, man and God. With Cusa and Galileo the middle term disappeared, and a yawning abyss was felt to separate the material from the mathematical, the real from the ideal, and man from his Maker.

A Body Called Light

The Grimms' tale of the "Bremen Town Musicians" ends with a group of frightened thieves cautiously reentering their forest hideout in the black of night. They have already been driven out once by four intrepid animal musicians. The lead thief tiptoes across the dark room toward the fireplace, where he sees two glowing orbs, which he takes to be hot embers. They are, in fact, the radiant eyes of the cat who shrieks and claws as he reaches for her, at which point donkey, dog, and rooster join in the fray to drive the intruders away.

The glow of a cat's eye, and her ability to navigate at night, convinced early opticians of the reality of the visual fire within the eye.* Yet the question still remained, if the antique mind thought a source of light was in the eye, why can't you see at night? Various answers were given, but Aristotle's was influential and straightforward: dark air was opaque. Light the lamp

---

*We now understand the phenomenon as the reflection of dim light in the spherical eye of the cat, and make use of the effect by putting tiny glass beads in the lettering of traffic signs to reflect the light of oncoming cars.

and it becomes transparent. Light for Aristotle was "the actualization of the potentially transparent."

To explain: modern calculators or portable computers often have LCDs or "liquid crystal displays." They work by making segments of the display selectively opaque. It is done by sandwiching a special "liquid crystal" between transparent electrodes and applying small voltages to those regions of the electrodes that should be changed from clear to black, from transparent to opaque.[24] Here the action of electricity changes the *state* of the liquid from opaque to transparent, or as Aristotle might say, from potentially transparent to actually transparent. In just the same way, fire can change the state of air from dark to "light," from potentially transparent to actually transparent. Then you, with your active eyes, see through the room. For Aristotle, light was not a thing but a condition or state of a medium. Aristotle's was a disembodied understanding of light, a subtle conception that easily accounted for the invisibility of light. Light had no substance or structure, but by the seventeenth century matters would change.

—

IN 1611, GALILEO had brought with him to Rome not only his famous telescope, but also a little box containing remarkable stone fragments of a mineral called by its Bolognese discoverer *spongia solis*, or solar sponge.[25] Placed in a dark room, these *cold* stony fragments would continue to glow if they had recently been exposed to light. Cold light! According to Aristotle, fire rendered air transparent, not cold stones. Perhaps the philosopher was wrong, and light, like heat, was better understood as a corpuscular and mechanical action. Galileo, therefore, suggested that when substance is reduced to "truly indivisible atoms" then "light is created."[26]

Light may be a body like all others but only smaller, or perhaps even the smallest. Perfect, incorruptible light, the sign, symbol, and very nature of God or Ahura Mazda, is really only another earthly body. Is such a suggestion any less revolutionary than the displacement of perfection from a spiritual universe to a mental abstraction? From the Milky Way (newly explained by Galileo) to the tiniest atoms of light, everything is uniformly material and, as a consequence, equally amenable to the new methods of scientific investigation being developed. Reason and experiment are now adequate to encompass the two infinities of the macro- and microscopic.

Galileo hesitated to give an unequivocal opinion on the nature of light, and even declared emphatically that he had "always been in the dark" on the question of "the essence of light" but a scant two years before his death. After reading an attack on his materialistic theory of light, Galileo rejoined with a story. He was willing to be imprisoned in a pitch-dark cell, he said, surviving on bread and water, if only he would be guaranteed on his release into light that he would know what it really was. If his sentiments were genuine, then he died unrequited, longing still to know the true nature of light.

Galileo had, however, voiced an opinion, one that gained intellectual momentum and conceptual detail. Light was not God but a body. If a body, then it would have an anatomy open to dissection and scientific investigation like other bodies. None achieved more in this vein than Sir Isaac Newton.

Newton: Standing on the Shoulders of Giants

We are like dwarfs sitting on the shoulders of giants; hence we can see more and further than they, yet not by reason of the keenness of our vision, but because we have been raised aloft and are being carried by men of huge stature.
Bernard of Chartres, 1115

Isaac Newton's fame began like Galileo's with the construction of an instrument, the same kind of instrument, a telescope in 1669.[27] As holder of the Lucasian chair in mathematics at Cambridge University, Newton, then in his late twenties, could do more or less what he had a mind to. And his mind chose to struggle with an array of questions that ranged from alchemy and theology to mathematics and optics. Out of his reflections, readings, and experimentation with optical phenomena, came the idea for a novel kind of telescope, one that did not rely on glass lenses* but on a curved mirror for gathering and concentrating light from distant objects. Laboring in his own workshop, the Lucasian professor ground a suitable mirror surface and made the tube and mount for what would become the prototype for all major astronomical telescopes the world over to this day. He took pride in the accomplishment of his own hands, and must have relished the initial failure of professional instrument makers who were later charged with replicating his design.

Except for a few colleagues at Cambridge, Newton's remarkable talents and diverse accomplishments were unknown. Rumors of his device eventually reached the recently formed "Royal Society for the Improvement of Natural Knowledge," who naturally requested the pleasure of examining it. Newton obliged by sending them a twenty-five-inch-long version of his reflecting telescope. After Society members studied it, they were unanimous and generous in their praise, longing to know more of Newton and to have him as one of their own. In response to their accolades, Newton sent his famous letter of February 6, 1672, to Oldenburg, then secretary of the Royal Society. In it he laid out the results of his experiments on the nature of light and color, and put down the foundation for the understanding that dominates popular discussions of light to this day.

*Refracting telescopes of the period suffered from "chromatic aberration," that is, images of differing color focused in different planes.

The origins of Newton's conception of light began early, with his boyhood love of tinkering and his natural inclination toward a mechanical philosophy of nature, but we will pick it up in 1665 when it "pleased Almighty God in his just severity to visit the town of Cambridge with the plague of pestilence."[28] During the two years of plague in Cambridge, Newton lived quietly at home with his mother in Woolsthorpe. These two years are legendary in the history of science, sometimes called the *anni mirabiles*, because of the sheer number and significance of the scientific ideas that came to Newton during his months of retreat. Certainly his previous years of energetic study had created the requisite fertile bed, but the scale of originality he demonstrated during these two years of meditation was unprecedented. Among the accomplishments Newton credited to those years (and there were others) were the invention of calculus, his theory of gravity and planetary dynamics, and his theory of light and color. Any one of these would have guaranteed his name to posterity.

What connected such disparate subjects as these, and so informed his and now our understanding of light, was a fundamental principle of which Newton was supreme master—analysis. It was not original with him, but he applied it with an unprecedented practicality and brilliance. As an example of the power of his analysis, let's look at an important derivation from his gravitational theory.

All material objects attract one another, but exactly how? How strongly and in what way does attraction depend on distance or the sizes of the objects? Newton solved the problem twice, once for himself and once for us. In the year 1666 while at home, Newton was musing in a garden on the problems of motion. As we all know, an apple fell. Yet Newton saw in its motion not what he and others had seen before. In the descent of the apple he also saw the travels of the moon. They were the same! If gravity could extend to the moon (a new thought), then the moon, too, should be falling. If, in addition, it was moving laterally as it fell, it might just go around the earth instead of hitting its surface. "Whereupon," we are told in early accounts, "he fell a-calculating what would be the effect of that supposition."[29] His calculations showed that his supposition might in fact work.

Twenty years later, in his *Principia*, Newton needed to convince us, his astonished readers, that he had not been hallucinating when he saw the apple and the moon as one, and for this he needed analysis. Ironically, Newton recognized that although a derivation of the universal law of gravitation for large bodies such as the earth could be done easily using his newly invented calculus, he chose to use the ancient and trusted methods of geometry, but adapted to modern requirements. These requirements were none other than those implied by the calculus where the ratios used may well be between quantities that approach zero. As Newton's major biographer Richard S. Westfall writes, "Euclid would not have recognized his offspring."[30] With Newton, the principle of analysis became formally embodied in mathematics and so, too, in his thinking. In that moment a fundamentally new kind of thinking, and so, too, of seeing, entered the West, one that challenged the human imagination to attain new heights of abstraction.

Until the seventeenth century, mathematics had been dominated by plane geometry. Even though the elements of geometry, such as lines, triangles, etc., are purely ideal and nonphysical in character, they remained visualizable. In neoplatonic thought, geometry dwelt midway between the material sense world and the pure unextended, formless world of Ideas. Indian philosophy made a similar distinction. Above the physical level was the *rupa* or "form" realm; beyond it was the formless, *arupa* realm. With the development of calculus, mathematics crossed to the realm of *arupa*, to the arena of hitherto unimaginable quantities.[31]

As mathematicians such as Newton and Leibniz were conceiving of infinitesimal quantities, they and others were simultaneously groping toward a material atomism of the kind hinted at by Galileo. The earth is an enormous mass, but it can be thought of as the nearly infinite sum of tiny masses. This is the key idea behind several of Newton's demonstrations in the *Principia*. The earth as a whole will attract the moon and the apple in identical ways because every atom of earth has its own unique gravitational relationship to the moon and apple. The total effect is gotten by simple addition (or integration in the continuous case).

Can analysis, so successful in physical dynamics, be applied to light? What are the least parts of which light is made? Can they be added together and separated not only in the mind but in the laboratory as well? Newton again provided the answer and showed others the means.[32]

—

IN LATER YEARS Newton was painted again and again as standing in a darkened room with a narrow beam of light passing from a shuttered window to a prism held before his calm but penetrating eyes. Exiting the prism was a rainbow of colors that brightened the air, falling on the far wall of the room. Here was the analysis of light, the extraordinary moment seen by Newton as the breaking up of light into its "least Parts."

In keeping with the artist's image of him, Newton opens his *Opticks* with a definition of the fundamental unit of light out of which the entirety of his optics would systematically be built. "Definition I: By Rays of Light I understand its least Parts, and those as well Successive in the same Lines, as Contemporary in several Lines."[33] The "ray of light" is the fundamental unit or conceptual atom of his theory. The origin of this definition is certainly Newton's early and abiding corpuscular conception of light.[34] However, the analysis of light occurs on two levels, one mental and the other physical, and when under attack,

Newton made good use of the distinction. In such times he invariably responded that his "least Parts" or rays were purely formal, theoretical constructions that did not commit him to a particular physical model of light.[35]

Light rays, according to Newton, are created in the sun and travel through space to us unchanged by reflection, scattering, or refraction. Moreover, each kind of ray produces in the eye a different sensation: red, green, blue, and so on. Natural sunlight is the sum of many such rays and so appears white. A prism acting on white light is the analyzing instrument that separates its constituent rays into their original classes. If the first prism is followed by a second, the "colorific rays" can be brought together again and so re-create white light. In Newton's hands white light, like the earth, was analyzed into its "least Parts," the various color-producing rays, and the prism was the device that performed the physical analysis.

Although reasonably careful to hide his particular model of light behind adroit philosophical language, Newton could not resist formulating his views on the nature of light, at least in the form of questions.[36] For example, in query 29 of the *Opticks* he asks, "Are not the rays of light very small bodies emitted from shining substances?" The smallest such bodies he thought would evoke violet and blue impressions with larger and larger corpuscles correspondingly responsible for green, yellow, orange, and red. Our sensations of color, therefore, were to be understood as our subjective response to the objective reality of corpuscular size.

In addition to suggesting that lights of various colors might ultimately be small bodies of various sizes, Newton explained important optical phenomena by calculating the change in trajectory of such bodies, for example, as they passed from air to water and were refracted. These ideas were drawn directly from his study of mechanics. The *Principia*, his book on mechanics, even contained a derivation of the law of refraction according to the corpuscular model of light.

Thus, not only was *light a body,* but its laws of motion were none other than the laws Newton had already discovered as those governing the motions of planets and apples. Forces of attraction and repulsion pulled and pushed projectiles of light through the world. Left undisturbed, they would travel in straight lines according to the law of inertia like all other material objects. The *dynamics of light* were identical to the dynamics of the planets.

The cosmos was unified, from phenomena on the grandest scale to those on the most minute, from the stars to the particles of light they emitted. It was a single vision. Gone was the multiplicity of being in which a unique intelligence was associated with each heavenly sphere. All was reduced to matter moving in obedience to the laws Newton unearthed. It was an impressive, even a compelling vision.

Newton, conscious of his accomplishments, feigned modesty by reaching back to a medieval image. In Chartres cathedral above the south portal rise beautiful, large stained-glass lancet windows. They show the New Testament Evangelists seated on the shoulders of Old Testament prophets, symbolizing thereby the foundation on which they stood as they continued God's work. Their vision rested on the past vision of the prophets. Newton's accomplishments likewise rested on changes in consciousness wrought by his predecessors. Correctly, he wrote: "If I have seen further, it is by standing on the shoulders of Giants,"[37] adding to their views his own powerful vision of the world.

—

ASIDE FROM THE objections of a few lone scientists, theologians, and artists, the response to Newton's achievement was euphoria. Scientists and philosophers of the day were dazzled, and contemporary poets and artists likewise joined in lavish praise of Newton.[38] For many years, some had felt the traditional sources of poetic vision were failing; they were dead or dying. The Muses no longer sang as they once did. The "twilight of the gods" had come, and like the fairy beings of legends, the gods had retreated before the crass onslaught to more congenial climes. If a poet would speak truth, where could he turn for inspiration? Finally an answer could be given—to Newton. John Hughes envisioned him this way:

> The great Columbus of the skies I know!
> 'Tis Newton's soul, that daily travels here
> In search of knowledge for mankind below.
>   O stay, thou happy Spirit, stay,
> And lead me on thro' all the unbeaten Wilds of Day.[39]

Newton became the inspiring spirit, the Muse of poet and philosopher alike, and many were the songs sung in his praise or verses written to make poetry of his scientific tracts.[40]

It would be a century before poets would turn on Newton and the despotism of his botanizing eye. They would then lament the dismembering of the world into parts, so that true wholes were nevermore seen. But for now most were content.

In the years following the 1704 publication of the *Opticks*, Newton's theory of optics, shorn of its subtleties, was broadcast to the widest audiences. Three different means were used. At the universities, his theory was usually accepted uncritically and without philosophical sophistication, as the basis for instruction. Since the truth had been discovered, there seemed little need for original optical research, and so none was undertaken. The competing wave theory of the Dutch physicist Christian Huygens was lost in the blaze of enthusiasm for Newton's optics, and for Newton himself. Not surprisingly, therefore, little criticism arose from university quarters.

For the public, another disseminator of Newton's thought was the popular traveling lecture-demonstrator who visited science clubs for both the upper and middle classes, demonstrating and explaining the latest seemingly magical, scientific developments. The third means was through popular accounts written by literary figures. Perhaps the most illustrious to undertake this task was the brilliant wit of the French Enlightenment, Voltaire. Tutored by his equally brilliant mistress, the "immortal Emily," whose love of mathematics and mathematicians was legendary, Voltaire sought "to remove the thorns from Newton's writings without loading them with flowers which do not suit them." Following the publication of his *Elements of Newton's Philosophy*, even his Jesuit adversaries had to admit that "all Paris resounds with Newton, all Paris stammers Newton, all Paris studies and learns Newton."[41]

—

THE FORCE OF these many efforts shaped a new way of seeing in the widest literate circles. The view of light gradually became sharper and more popular. What had at first been the province of a few mathematically minded scientists was taken up again and again by artist and author and academic. What their renderings sacrificed in philosophical subtlety was replaced by materialistic simplicity. Gone were Newton's or Galileo's hesitations about the "true" nature of light. Light was a body, and its movements were like those of all other bodies. Such were the facts.

In France, a parallel, if distinctive, line of inquiry had been launched by René Descartes some decades earlier. Like Newton's understanding of light, it was founded on a rational and predominantly material conception of the universe. Cartesian and Newtonian physics vied for dominance for more than a century. Ironically, the origins of Descartes's universal mathematical science undercut his rationalist agenda, for Descartes received his mandate in the classical manner of all biblical figures, through visitation in a dream.

Descartes's Dream

I frankly confess that in respect to corporeal things I know of no other matter than that which the geometers entitle quantity.

Descartes, 1644

While at Cambridge, the twenty-two-year-old Newton purchased a copy of René Descartes's book *Geometry*. He read the first two or three pages and found he could understand no more. He returned to the beginning and started again, managing to comprehend another three or four pages, and so it continued until he had satisfied himself concerning the mathematics of this French master. Here was a book and a thinker worth troubling over.

Forty-six years Newton's senior, Descartes had advanced a philosophy of nature in which the universe was a mechanism, from the simplest atom to the most complex aspect of the living human anatomy. "The rules of nature are the rules of mechanics," he declared.[42] This dictum was taken up and promoted in the most powerful ways by his friend and advocate, the commanding French monk Marin Mersenne. The "mechanical philosophy" of the seventeenth century was largely forged by these two Jesuit-trained giants, rising in Descartes as the fruits of his solitary meditations, and then promulgated by Mersenne through his vast correspondence and highly influential connections. At their hands, as well as under the swords of the Duke of Bavaria's Catholic armies during the Thirty Years War, the magical philosophy of Renaissance animism, Hermeticism, Cabalism, and all their attendant trappings, were destroyed. In their place would emerge Descartes's "universal and admirable science." How deliciously ironic that the conception of Descartes's consummately rational science occurred in a dream.

In November of 1619 Descartes, a twenty-three-year-old soldier-philosopher, retired for the winter to a modest house near Ulm.[43] During the previous twenty months, he had traveled through Germany and Holland, corresponding with mathematicians and philosophers whose range of interests included not only the orthodox problems of science, but also the connections between science and spirituality. Some were certainly involved in Rosicrucian spiritual pursuits, which in these years were experiencing an enormous surge in public attention.[44] Two short tracts concerning Christian Rosenkreutz and his brotherhood had just been published, the *Fama fraternitatis* (1614) and the *Confessio* (1615), causing the so-called "Rosicrucian Furor." The *Confessio* put forward the tenets of the fraternity of Christian Rosenkreutz, which included the conviction that on the basis of God's revelation and human invention, observation, and understanding, "if all books should perish, and by God's almighty sufferance, all writings and all learning should be lost, yet posterity will be able only thereby to lay a new foundation, and bring truth to light again." This was to Descartes's liking. He searched for a member of the "invisible brotherhood" but maintained he had found none. Through his contacts with other seekers, however, the impressionable Frenchman was infused with the holy significance of his task: the creation of a new science of nature.

In the middle of these influences, and following a period of intense solitary meditations, Descartes tells us that on November 10 he had a vision and a three-part dream that revealed to him his vocation and the foundations of the science he was to create. He held fervently that this single episode was the most important occurrence in his entire life, and in gratitude for it he vowed to make a pilgrimage of thanksgiving to the Virgin at Lorette, a promise he kept five years later when he traveled there on foot from Venice. Descartes's own detailed account of his vision and dream has unfortunately been lost, but we have a faithful, if incomplete and pale, summary of it from his early biographer,

Baillet, who studied it. To our unholy eyes it may seem unremarkable, but Descartes knew it to be his personal Pentecostal visitation, his life's epiphany.

In the first part of his dream, Descartes finds himself struggling against a tempestuous wind in order to reach the Church of the College of La Flèche (where he and Mersenne had been educated) to say his prayers. Descartes turns to show courtesy to a man he has neglected to greet, and in that moment is blown violently against the church. He is presently informed that someone has something for him—a melon. Descartes awakens to the experience of pain, turns over on his right side, and prays for protection. Falling asleep once more, he has another dream. Concerning it we only know that it fills him with terror; he is awakened by a noise like the sound of thunder and sees thousands of shimmering sparks in his room. In the final part of his dream, Descartes sees a dictionary and a *Corpus poetarum* on his table, open at a passage of Ausonius: *quod vitae sectabo iter?*, which means "what path shall I follow in life?" An unknown man appears and hands him a bit of verse on which Descartes reads the Latin phrase, *est et non*, "yes and no."

We possess, through Baillet, but a few meager bits of Descartes's own interpretation of the momentous night. Significantly, the lightning was taken by Descartes to be "the Spirit of Truth that descended upon him and took possession of him." The dictionary symbolized all the various sciences, and the *Corpus poetarum* "marks particularly and in very distinct manner, Philosophy and Wisdom linked together."

Thus, we find the youthful Descartes in his solitary reflections in Germany, searching for a means to truth. He has made contact with the forbidden sciences and is deep in meditations on his future life. Descartes tells us that "the genius that heightened in him the enthusiasm which had been burning within him for several days past, had forecast these dreams to him before he had retired to his bed." Thus Descartes knew that the night of

November 10 would bring the answer to his passionate quest. What was the answer he heard? It was, as Jacques Maritain calls it, the Pentecost of Reason.

In the months following, Descartes would write his most important foundational work, *Discourse on Method,* whose original title was to have been *Project of a Universal Science Destined to Raise Our Nature to Its Highest Degree of Perfection*. As the *Confessio* had said, all the prior works of scholars needed to be set aside. Until now we have "been governed by our appetites and our preceptors. As a result of these reflections, it was borne in upon me [Descartes] that as far as all the opinions I had received thus far were concerned, I could not do better than to undertake once and for all to get rid of them." The Spirit of Truth had descended on him; it was his solitary task to adjust everything to "the level of reason." His admirable and universal science was not the work of a brotherhood, occult or public, but the creation of a single mind, his own. Descartes's was to be a godly science that rises with certainty to the simple intuited nature of things as God knows them. Maritain has called this science "the mythology of modern times that promised everything and denied everything, that raised above all things the absolute independence, the divine *aseity* of the human mind." It was the child of a mischievous genius conceived in a philosopher's brain—the Dream of Descartes. Within this dream, what is light?

—

SPACE, DESCARTES ARGUES, is inconceivable apart from matter. Therefore, where there is space or "extension" there must also be matter, always. In an analogy that befits his nationality, Descartes likened our universe of space to the winemaker's vat just after the harvest. The vessel is filled with half-crushed grapes, which are completely surrounded, even permeated, with the juices they have given up. Likewise, the vast reaches of space carry planets, stars, and moons like grapes in their own juices. Space for Descartes is filled with an atomistically conceived *plenum*, a material fluid that occupies all voids and drives the planets in their courses like bits of grass caught in the whirlpools of a stream. Descartes could thus command, "Give me motion and extension and I will construct the universe!"[45] And light?

In Descartes's view, between the eye and every object is a column of *plenum* along which an action can travel. Light is neither a projectile nor a fluid flow, but a "tendency towards motion" in the *plenum* that propagates at infinite velocity along the column. Sight, he writes, is like the blind man's stick. As he walks he feels around him with the stick. An object that strikes one end instantly causes a push at the other. In the same way, an object affects the *plenum* around it, causing a concussion at the eye, and so we see. Sight and light, according to Descartes, were to be understood as pure mechanism. This is a watershed analysis that offered an imagination of nature and a basis for scientific inquiry that would dominate science for three hundred years. As the distinguished Harvard historian of science A. I. Sabra writes, mechanical analogies had been used long before Descartes to help explain certain optical phenomena, "but the Cartesian theory was the first clearly to assert that light itself was nothing but a mechanical property of the luminous object and of the transmitting medium. It is for this reason that we may regard Descartes's theory of light as the legitimate starting point of modern physical optics."[46]

While the impact of his conception of nature and science proved seminal, the specifics of Descartes's theory of light were relatively short-lived. Across the waters of the channel, Newton drew out certain of the absurdities to which such a view of light and cosmos led. By contrast, Newton's own dynamical formulation of optics along mechanical lines gained widespread acceptance, but it, too, suffered from serious deficiencies. In the midst of the triumphant march of Newton's corpuscular optics, persistent critical voices could be heard. Many spoke of specific failings of the theory.[47] For example, an analysis of two people staring at one another in the projectile view would seem to require that particles of light pass simultaneously along the same path in opposite directions. Also, if the sun was giving off copious numbers of particles, why does it not waste away over time? Proponents replied that the corpuscles are extremely small, so that the sun loses but a drop or two of substance in the course of a day. Or what of the innumerable corpuscles that must pass through the pinhole of a *camera obscura* without affecting the image? Again smallness came to the rescue. Yet if they are so small, then how can they be so effective? When investigating the impact of intense, focused light, the experimentalist John Michell had been troubled by the melting of his copper-plate detector at the focus of a two-foot reflector. Light can yield dramatic effects. If the bodies of light are small, they must also be exceedingly numerous in order to account for such power.

Nieuwentijdt in his book of 1718, *The Religious Philosopher*, calculated that 4.1866×10^{44} particles are emitted each second from a candle flame.[48] This approaches the number of protons in the entire earth. Nieuwentijdt took the calculation as evidence for the constant attention of God to his creation, the "Direction of an Omnipresent Power, extending its Care over all things, even the smallest of Bodies." If light corpuscles were immutable as Newton claimed, why does the solar sponge seem to emit light of a different color from that which originally illuminated it? Such were the unresolved issues.

Descartes's theory avoided such difficulties by declaring light to be only a "tendency towards motion," not entailing actual motion. In this way Descartes tried to avoid the absurdity of having an object (corpuscles of light) move simultaneously in two opposite directions. The particles of the *plenum* do not actually move, but only "tend" to move. Rays of light are, therefore, "nothing else but the line along which this action tends."[49] Light is not, according to Descartes, the flight of a projectile, but the propagation of that action.

Hidden within Descartes's obscure conception of light is a strength. As puzzling as his view appears, it holds the seed of an enormously fruitful conception of light, the so-called wave theory, that would not only overthrow Newton's understanding of light, but would ultimately work against Descartes's own dream of a mechanical universe.

While there were public and poetic accolades for the corpuscular theory, scientists of stature like Huygens questioned it along the above lines. They put forward alternative conceptions of light. Some spoke of light as a material fluid of fire that flowed from the fountain of the sun and back again. Others filled the universe with a substance far more tenuous than air but of unbelievable rigidity whose vibrations were light. All shared a material conception of light, but the most successful theories also possessed elements that would, in the end, undermine that very position.

———

SWIM OUT INTO the ocean and notice a few things. The swells of the open sea slowly increase in size as they approach the shore. As they speed toward you there is a moment, especially for the inexperienced, in which you worry that you will be carried along and dashed to pieces on the beach. As the wave and the feeling pass, you notice that you and the water around you did not travel with the wave toward the beach, but only bobbed up and down. No water went by, it only moved up and down. What did pass? You saw it coming, and see it still heading for the beach. What is it, this wave? It is a form, a shape, a specific action of the sea.

Could light, like sound, be a shape, a vibration of a universal ether? Nothing moves, or moves only slightly, except the form and figure of light. Light then would not be substance, but rather form! With these thoughts, a small but serious competitor arose to the projectile theory of light, one that would gradually gain momentum and find a surprising ally in the new investigations then under way in electricity.

—

WHAT HAD BEGUN as a god, and in Greek hands had become a luminous inner fire whose ethereal effluence brought sight, became in the Middle Ages "the first corporeal form" of Grosseteste. His was still a metaphysics of light, but within that first corporeal form lay the seeds of light's final corporeal form: Newton's corpuscular optics. Along the way, perfection was deposed from its ancient haunts, and in its place reigned abstract mathematics in brilliant formality. Abstract perfection and substantial reality, these replaced the moral perfection and power of immortal gods. This was far more than an exchanging of ideas; it entailed a profound transformation in the West's very way of perceiving. A material and mechanical eye replaced the moral and spiritual one of earlier times.

With the scientific revolution of the sixteenth and seventeenth centuries, mankind entered a new age. It saw itself as entering manhood and so strove to put away childish things. Light was shorn of its metaphysical dress, its body was displayed naked like one of Leonardo's cadavers. Light's hard skeleton was laid bare by the incisive minds of Descartes and Newton, its composition and principles of motion exposed to enlightened scientific sight. As Bernard de Fontenelle, secretary to the Paris Academy of Sciences, wrote in 1686, the world was but a staged performance, an opera. The attentions of science were not on the dramatic action no matter how lavishly set or passionately enacted, but rather with the calculating view held by one usually unnoticed during the production—the stage engineer.[50]

The opera-goer does not care how the stage machinery works. By contrast the engineer in the pit hears every cue, sees the garish makeup of the actors, operates the lighting, raises the curtains, and sets in motion the mechanisms of the stage. The performers and technicians conspire to create an illusion, one that entertains a contented audience, but the engineer alone knows the full ungarnished truth.

According to Fontenelle, nature conspires in a similar manner to present a beautiful pageant to the human senses. We can be entertained by a sunset, a bird's song, or stirred by human passions to love and war, but behind these external happenings lies nature's secret machinery, the means by which she contrives to present her illusions to an unwitting audience. Unlike the common man, the scientist is not satisfied with the show, with seeing only the surface phenomena. Rather he relentlessly searches for the real mechanical operations of nature. As Fontenelle wrote, "He that would see nature as she truly is must stand behind the scenes of the opera."

Fontenelle, like Voltaire, wrote in order to enlighten the curious public about the mechanisms scientists were imagining as operative behind the outer display of nature. From his pen, as well as that of Mersenne, ran the ink that would enamor the public of a mechanical cosmos. The dream of Descartes would become the dream of an age. The enthusiasms of youth, however, are often contradicted by the lessons and puzzles of a long life. Such was the case also with the history of light and the mechanical universe. Light kept true to its own inviolable character, always ready to cast a new shell of curious design upon the shore of human imagination.

CHAPTER 5

The Singing Flame: Light as Ethereal Wave

We shape the clay into a pot, but it is the emptiness inside that holds whatever we want.

Lao-tzu

Throughout history, space has been like the emptiness inside Lao-tzu's clay pot. It's not the pot, but the emptiness inside that holds whatever we want. Ever since we created the concept, space has held whatever we put into it. We have imagined space to be many things, and that act of imagination has had implications for our image of light. Endow space with divinity and light is godlike; discover its shape and light is geometrical; fill it with matter and light is substantial. From Moses to Einstein, the history of light is also the history of space.

When on the first day of creation the Lord said, "Let there be light," the early Christian fathers understood the first light to be that noble, spiritual reality they termed *lux*, which was the soul of space. They, and medieval scholars after them, labored long and hard to distinguish *lux* from its emanation or bodily counterpart, called *lumen*. The distinction for us is difficult to grasp, yet was essential to their view of the world.[1]

Lux was God-given, essential light, the being of light, and as such a reflection of its Maker. Augustine viewed it as the simplest, noblest, most mobile and diverse of all corporeal being. *Lumen*, by contrast, was the material means by which our perception of the being of light (as *lux*) arose. When we sense the brilliance of the sun, we are perceiving its *lux*, but we do so by means of the unseen *lumen* that connect it to us. Between the time of Augustine and Galileo, the being of light that ensouled space (*lux*) retreated, leaving behind its hard material vestige (*lumen*) as a fossil record for the curious natural philosopher.

Light has fascinated and continues to fascinate, for, as Leonardo wrote, "Among the studies of natural causes and laws, it is light that most delights its students."[2] From the radiant eye of ancient Egypt to the quantum-field theories of today, light has sculpted space to suit its demands.

In Grosseteste's imagination, the unfolding of light from its primordial hearth gave birth to space as light multiplied itself, generation after generation, until, spent, it died at the shoreline of the universe it had created.

The space of Euclid and Brunelleschi was pure geometry. Light and vision proceeded as rays, a pure geometry of sentient lines linking soul to world.

For Descartes, space was dimensioned, extended, and therefore had to be substantial. He could not conceive of an extended space apart from substance; where there was one, there must be the other. Looking up by night or by day, one sees "a sky that is made of liquid matter," and the planets swirl in its whirlpools like grass in the eddies of a stream.[3] Light, whatever it is, must transit this material medium. By the end of the eighteenth century, this material medium was the ether whose motions became the *lumen* that caused vision, and *lux* was no longer an attribute of God but merely a subjective phantom of the mind.

Our understandings of light are thus entwined with our conceptions of space. They have coevolved: moral space and spiritual light, perspective space and geometrical light, material space and substantial light. Each age emphasized one face of light and so revealed its own predilections. The corpuscular conception of light emphasized its substantial nature, and yet it suffered from serious deficiencies. Perhaps, as some suggested, rather than consider light to be a material substance or fluid, it is a pure form, a figure dancing. After all, the mysterious nature of sound has finally succumbed and shown itself to be the vibration of air. Might not light profit from a similar treatment? It seems appropriate that a mathematician, whose life's work is constant contemplation of the insubstantial, should first advocate light as a figure dancing through the ether.

At fifty-three years of age, and nearly blind, Leonhard Euler, the greatest mathematician of the eighteenth century, the author of innumerable mathematical and physical treatises, and the most prolific member of both the Prussian and Russian academies of science, honored the petition of a curious young German princess from Anhalt-Dessau to hear his thoughts on science. His letters to the princess between 1760 and 1762 covered every aspect of science and quickly captured the attention of Europe, with thirty-six editions in nine languages. Having written the princess concerning sunlight, he anticipated her question, and ours. "Having spoken of the rays of the sun, which are the focus of all the heat and light that we enjoy, you will undoubtedly ask, What are these rays? This is, beyond question, one of the most important inquiries in physics."[4]

His response, coming as it did in an age reverberating with the ideas of Newton, was a piece of heresy. The rays of sunlight are, he rejoined, "with respect to the ether, what sound is with respect to air," and the sun is but a bell ringing out light.[5] With these words the first significant challenge to the Newtonian conception of light was sounded. The princess's reply: "Yes, but what, sir, is sound, and what the ether?" If we would understand light by analogy with sound, then we must first have some certainty about the nature of the latter.

Knowledge concerning sound did not come quickly. Its spiritual origins were enveloped in the nimbus of the Word, its investigators included the masters of all times. Yet haltingly, certainty arose concerning it, and by Euler's time a full response could be given to the princess. We, like he, must deviate from our tale of light to explore for a while the history of sound. Once we understand sound, we will appreciate the graceful analogy Euler suggested for the nature of light.

Of Harmonies and the Vacuum

From the magnificent north spire of Chartres cathedral, the "Sun Tower," bells call the faithful to worship. Three hundred feet below, in the archivolts of the Royal Portal, are other, silent bells of stone. They appear as part of the twelfth century's portrayal of a learned Christian life, a scholarly and meditative life that moved through the mastery of seven "liberal arts." These seven disciplines formed what William of Conches called "the proper and only instrument of all philosophy." Four of them were devoted to the sciences. These four, the Quadrivium, "enlightened the mind," and one was the science of music.[6] The medieval patron of music shown in the sculptural depiction at Chartres is not an early Christian composer but the pagan philosopher Pythagoras, who sits working while listening to the ringing of bells and the tones of a lyre, instruments by which he first discovered the presence of number in all things. It may seem that he is listening to earthly tones alone, but the Chartres masters knew well that he listened also to the inaudible concord

Pythagoras at Chartres cathedral listening to the harmony of the spheres.

of heavenly music, whose divine harmonies echoed throughout science from Pythagoras to Johannes Kepler. Sound, like light, has a spiritual as well as a secular history, one whose trajectory parallels that of its visual companion. According to the Finnish folk epic, the Kalevala, the world was sung into existence. Before sound could become the referent for a mechanical image of light, as it was for Euler, it had to empty itself of its own spiritual nature, it had to become a body and not the eternal reverberation of the Word.

The transition to an entirely mechanical image of sound took place in the seventeenth century. A characteristic figure in the story was the learned Jesuit Athanasius Kircher, who combined a religious temperament with a modern bent toward experimental inquiry and debunking pagan superstitions. For example, on the one hand, he understood Job's line "When the morning stars sang together . . ." as a reference to the Pythagorean harmony of the heavenly spheres, a subject on which he expounded at some length in many books.[7] Yet while longing to hear the Lord's pure song, Kircher tinkered in his laboratory and museum of sound at the Roman College devising novel acoustical instruments that he often used to disabuse the commoners of their superstitions. For instance, on Pentecost, that religious feast set aside to commemorate the miracle of the Holy Spirit in which tongues of flame descended on the first Christian disciples, Kircher and his companions dragged enormous speaking trumpets to the top of St. Eustachius mountain. Acting the part of an angelic choir, they sang litanies that could be heard in villages up to five miles away. Over two thousand villagers followed the call from above, bearing gifts to these brazen angels.

Yet Kircher was also an earnest investigator whose most important scientific experiment was the first attempt to study the propagation of sound through a vacuum. The very idea of the vacuum, of space devoid of matter, was a hotly contested philosophical point in Kircher's day. Aristotle had long before declared that "Nature abhors a vacuum," and his words carried enormous weight. Notwithstanding the controversy, in 1650 Kircher inverted a column of mercury (as is commonly done in mercury barometers) and placed a small bell in the empty space above the mercury. Waving a lodestone—a magnet—by the enclosed iron clapper, he caused the bell to ring, and hearing it do so through the vacuum, he quite sensibly concluded that air was *not* necessary for the transmission of sound. Others in the Accademia del Cimento of Florence repeated the experiment with similar results and conclusions. All this seemed to support the view, held by the French philosopher and atomist Gassendi, that sound was caused by the emission of a fine stream of invisible particles that made their way from the source to the ear. Not until ten years after Kircher's experiment did Robert Boyle, using a much-improved, von Guericke vacuum pump, succeed in showing that the ringing of a bell could *not*, in fact, be heard if carefully suspended in an evacuated glass container. Air *was* necessary for the transmission of sound.

Interestingly, one could still see through the vacuum; it was not dark, meaning that light, by contrast, did not require air for its transport. Although light, unlike sound, did not require air in order to travel, it remained entirely possible that a far more subtle substance remained in the region, a substance that could not be eliminated by Boyle's vacuum pump.

The discovery that sound requires a material medium for its travel, and allied discoveries such as the precise measurement and prediction of the speed of sound, all succeeded in tying sound to the earth. In the scientific imagination, sound became a mechanical and material phenomenon. But the specific nature of sound, as Bacon wrote, remained "but superficially observed," and was "one of the subtlest pieces of nature."[8] How is sound physically produced, and what is its specific nature during its flight through the air?

Sing a few notes, and as you do so touch your larynx or put

your fingers firmly into your ears. The vibrations you feel hint strongly at the mechanism associated with the production of sound. Look closely at a bowed violin string, or the end of a mouth harp. Its blurred motion naturally suggests the association of vibration with the generation of sound, an association already made by Aristotle.[9] Yet how do such vibrations correspond to sound in detail?

The mathematical relationship between pitch and the tension or the length of a string goes back to Pythagoras, but the association of it with a specific frequency of vibration originates with Galileo at the end of the "First Day" of his *Discourses Concerning Two New Sciences* of 1638. Should you happen to scratch your fingernail on a blackboard, the squeal produced not only sends a shiver up your spine, but you also feel a vibration at your fingertip. Try it with a piece of hard chalk, and you will notice that the vibrating chalk leaves a trail of marks behind that are closer or farther apart depending on whether the pitch is high or low. This observation was first casually noted by Galileo when he ran a hard iron chisel on a brass plate. Then he repetitively ran his chisel over the brass plate under varying conditions. Although the screech must have been ear-shattering, the results were exhilarating. With each vibration the chisel left a mark in the plate. By counting the marks per inch, Galileo could correlate the pitch produced with the vibrations per second. He was even able to confirm the ancient Pythagorean ratio for the musical interval of the fifth by sounding his chisel-plate instrument, comparing it to a well-tuned clavichord. Pitch and vibrational frequency were thereby yoked together. Sound became a vibration that passed through the material medium of air.

Remember the image of the ocean wave moving toward the shore. The wave, rushing by, raises and lowers the water but does not drive it toward the beach. All wave motion is similar in this regard. A small, sometimes very small motion is enough to send a "shape" rippling quickly away from its source. The medium that carries that shape merely shakes in the tiniest way, yet a thunderclap travels eleven miles in a single minute and can be heard twenty miles away.

—

TWO HUNDRED YEARS after Galileo, during the performance of the grand trios of Beethoven, and long after it was well established, the vibrational basis of sound received unexpected visual confirmation. While listening to the string trio, the attention of John Leconte, M.D., wandered to two fishtail gas burners near the piano. The flame of one was seen to be pulsating in *exact synchrony* with the music.[10] Even the trills of the cello rippled beautifully across the sheet of flame so that "A deaf man might have seen the harmony." The hidden vibrations of sound were made visible by a "sensitive flame."

The cold, dim glow of his solar sponge had suggested to Galileo that light might be a body, the smallest in existence. Perhaps he was wrong, perhaps one could replace his dull material conception of light with a view of it as tiny waves rippling through an ether like the trills of a cello rippling across an open flame. Perhaps light is a singing flame, a delicate vibration of the luminiferous ether.

The Two Faces of Knowledge

Sound, shorn of its ancient associations with the creative Word of God, presented itself to the clear-eyed intellectuals of the eighteenth century as a wave of alternating compression and rarefaction in the invisible medium of air. With this view of sound established, it was natural to consider light to be of a similar nature. The imagination, scientific or otherwise, quite naturally sees the unknown in familiar idioms: God in the image of man, light in the image of sound.

Before going further, we should pause in order to realize the incompleteness of the picture we have produced of sound. Pure, unvarying vibration carries no meaning; this is a theorem in mathematical physics. Bow a B-flat on your violin *forever* and nothing has been conveyed, not music, not voice; no "signal" has been transmitted. In fact your sense of hearing knows this perfectly well and so will tend not to hear an unvarying background sound whether it be a waterfall or the hum of a badly ballasted fluorescent light. In order to speak we must modulate the pure tones we produce, forming them into words. Meaning arises as much through silence as through sound. Count Maurice Maeterlinck knew this when he wrote, "Souls are weighed in silence, as gold and silver are weighed in pure water, and the words which we pronounce have no meaning except through the silence in which they are bathed."[11]

Vibration of the air is like the unformed lump of clay before the hands of the sculptor have shaped it into the powerful creation that can so move us. The human larynx acts like hands, shaping the monotone of the physicist's audio oscillator into the diction of meaningful speech. Could the same be true of light? Adelard of Bath in the twelfth century wrote that sight was due to a "visible breath."[12] He thought that we "inhale" external light (*lumen*) and "exhale" the light of meaning (*lux*), which reminds one of the reciprocal action of outer and inner light written of by the Greeks.

To see sound or light as vibration only is to reduce Michelangelo's *David* to marble dust. One may in a sense be correct to do so, but in the process we lose the truth the statue embodies. The untouched marble is all potential. Like the god Proteus, it can assume any shape. Left to itself, it conveys nothing. To use the language of light, it is as if *lux* must set upon *lumen* so that speech, music, or a bird's gentle song can all touch us, physically and psychically. If vibration is the body of articulate tone, its spirit is reflected in its infinitely nuanced form. Sight like hearing requires a modulated and crafted form of light for meaning. Stabilize images perfectly on the retina and they disappear. This is a fact of sense psychology. We see only change, movement, life.

—

THE SPECIFICS OF speech and hearing were separated off into another discipline (the beginnings of sense physiology and psychology) during the eighteenth century, leaving the "bodily aspects" to physical scientists. Among them Leonhard Euler provided the first carefully supported vibrational theory of light in his *New Theory of Light and Color* of 1746. Luminous objects "vibrated," he wrote, and the ether carried those vibrations to the eye as air carried sound to the ear. To advance his own vibrational theory, he first needed to discredit Newton's corpuscular view, and this he did more systematically than anyone before him, using many of the objections I raised at the close of the previous chapter. Still, even Euler and his vibrational theory could not explain all of light's many manifestations, and primary among them were the phenomena of diffraction. The word "diffraction" was coined by the Jesuit Father Francesco Maria Grimaldi of Bologna, Italy, following his detailed investigation of light phenomena in 1665. Diffraction effects are subtle presences in our lives of which we are usually unaware.

Take from your wallet a credit card. It very likely possesses a mirrored patch in which an image hovers. Occasional magazine covers, promotional gimmicks, and every science museum display similar three-dimensional, "holographic" images. All are diffraction phenomena.

On a rainy night, look through your umbrella at a streetlight. The multiple, colorfully transformed image of the light you see is a diffraction phenomenon.

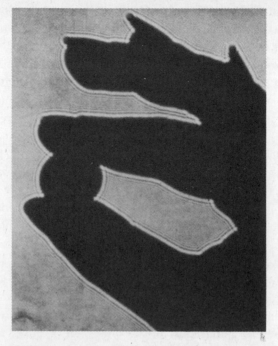

The diffraction of light shows up as fringes en-
circling the hand and coin. Here laser light is
used.

With your fingers very close together but not quite touching,
hold them up to your eye and look through the small gap toward
a source of light. The pattern of dark lines and shapes you see
is a diffraction effect.

Of the countless arrangements that give diffraction, the simplest is surely light passing by an opaque edge. At an unrelenting
boundary, free white light abruptly meets and succumbs to darkness. The offspring of this meeting are subtle but unmistakable.
Where before one noted only light and dark, variegated, parallel
bands of color rhythmically intrude into each region. Unnoticed
until the seventeenth century, the phenomenon enriches the

metaphor of light. Once again light struggles with darkness and the meeting gives rise to color.

None of these phenomena could be explained with a corpuscular view of light. Nor was Euler's wave theory adequate to the task, because it was still missing a key concept. If light was like sound, then such diffraction effects should appear in sound as well. At that time none was known. As it turns out, sound does indeed show such effects but not so readily. For example, sing a note in the bathroom. Slowly vary the pitch and notice the change in intensity. When the frequency of vibration is resonant with the room, the sound waves interact to produce a clear increase in volume. Even this simple acoustic phenomenon could not yet be understood. Something important was missing from the understanding of waves whether acoustical or luminous. The concept lacking was provided by the English scientist Young and masterfully applied by his contemporary Fresnel in France.

—

THOMAS YOUNG WAS a prodigy and polymath of amazing reach.[13] Already reading by age two, in his early years Young was largely self-taught in mathematics through the calculus, and the natural sciences (including the making of telescopes and microscopes). He displayed an early and abiding fascination with languages, studying first Latin, Greek, French, Italian, and then going on to Hebrew, Chaldean, Syriac, Samaritan, Arabic, Persian, Turkish, and Ethiopic. Young, in fact, successfully deciphered parts of the famous Rosetta Stone independently of Champollion and so helped to unravel the hieroglyphic script of ancient Egypt. In London, Edinburgh, and Göttingen, he studied medicine, receiving his M.D. in 1796. His interest in vision and his study of the eye led him to the seminal suggestion in 1801 that color vision occurs through the retina's sensitivity to three principal colors: red, yellow, and blue. Maxwell and Helmholtz later modified and extended Young's views on tricolor vision to the form now accepted today. Precocious, a lover of dancing and dressage, Young was an elegant and brilliant intellect whose importance for us lies in his revolutionary "principle of inter-ference."

Deeply influenced by Euler, Young subscribed to a vibrational theory of light and a universal luminiferous ether. In addition, however, he advanced a principle that he maintained could make diffraction phenomena understandable, which was a daring claim. Just as intersecting water waves may enhance or cancel each other, Young suggested that the undulations of the ether can be strengthened or weakened to the point of extinction through similar interference. We now are habituated to such ideas in high-school science classes, but we should not under-estimate the apparent insanity of Young's suggestion. According to this principle, parts of a uniformly illuminated screen can be made *dark* by the *addition* of more light. Light plus light equals darkness? This is just what Young was proposing.

Henry Brougham spoke for many contemporaries when he declared Young's principle of interference to be "one of the most incomprehensible suppositions that we remember to have met with in the history of human hypotheses."[14] Together with their rejection of the principle of interference, critics also rejected the hypothesis that light is the sensory effect of vibrations in a luminiferous ether. About the ether Lord Brougham wrote: "From such a dull invention nothing can be expected." Young, misunderstood time and again, came to see himself as a modern Cassandra who spoke nothing but truth but whom no one could understand. Still, the specters of diffraction and polarization left little real comfort for those who imagined light to be a stream of tiny projectiles, and the time would not be long in coming for the overthrow of their treasured hypothesis, and the vindi-cation of Young's incomprehensible principle of superposition (an incomprehensibility, by the way, that becomes all the greater in quantum mechanics).

—

THE TREATMENTS OF light by Newton, Descartes, Huygens, Young, and Euler differed in particulars but shared an important feature. They all stressed an analogical understanding of light. That is, light was viewed as like something else: a bit of matter, or the waves on a pond's surface. However, Euler's work also developed a parallel approach, namely the formal and mathematical description of nature.

Euler modeled light on sound, and so participated in the well-established tradition of proceeding by analogies with better understood phenomena. His mathematical temperament, however, led him to pursue an alternate description of phenomena as well, one that created a more abstract imagination of nature through the application of advanced mathematics including the newly developed methods of calculus.

Euler was certainly not the first to apply mathematics to nature. In fact, like the history of light and sound, the application of mathematics to nature has its own fascinating history that mirrors the evolution of human consciousness. Going back in time to ancient Babylonia, priest-astronomers stood atop their ziggurats or stepped pyramids to observe the movements of sun, moon, planets, and stars. On the basis of their observations, they fashioned a purely arithmetic astronomy of amazing precision.[15] Significantly, they completely lacked any picture of the cosmos other than that provided by religious mythology. In the intervening centuries, the consciousness of the astronomer changed profoundly. The planets and stars were no longer the dwelling places of the gods, but distant masses arranged geometrically around the earth or sun. Mathematics also changed. Its objects were not limited to the numbers and operations of arithmetic, nor to the elements and proofs of Euclidian geometry, but by the eighteenth century included radically new and seemingly unimaginable entities such as the infinitesimals of the calculus and, by the century's close, the first approach to non-Euclidean geometries. The perfection that Galileo drew down from the heavens and placed into mathematics quietly unfolded in sometimes disquieting ways.

Euler and his eighteenth-century contemporaries made great strides in the application of modern mathematical analysis to nature. Like the dance of sun and shadow about an object, Euler and every physicist after him would use two languages to discuss the nature of our world. One appeals to our sensual imagination, the other to abstract reasoning. From his day to ours, the being of light has been pursued using not only the nets of mechanical analogy, but also the far more subtle and insubstantial web of mathematics. Many raced to join Fontenelle, who watched nature's performance from the wings of the opera's stage where pulleys and props, makeup and lighting, could be scrutinized at close range. For a handful of others, however, the machinery of nature was interesting, yes, but not nearly so beautiful as the view of her forms and patterns to be seen through the lens of mathematics.

Euler addressed not only, nor even primarily, a German princess, but rather a different and more intimate audience in his hundreds of papers and books. He wrote these works for a tiny and elite community, one whose precarious existence depended on the whim and generosity of despots such as Catherine II or Frederick the Great. In the very same year in which his immensely popular *Letters to a German Princess* appeared, Euler published a learned volume entitled *Rigid Mechanics*, which sold all of twelve copies in its first two years. Amazingly, the spiritual progeny of those twelve readers, among whom would number Lagrange, Laplace, Poisson, Fourier, and Gauss, were the first mathematical physicists. They would, in the end, shape all of scientific culture. They, and not the German princess, labored hard to nurture the tender seedling of modern science. In the process, the very language of scientific communication, especially with the invention of the calculus, became less and less the common property of educated people. The widening abyss between the mathematician's intimidating discourse and popular wit is probably nowhere better illustrated than in the encounter between Euler and Denis Diderot, the brilliant architect of the great French *Encyclopedia,* at the court of Empress Catherine II of Russia.

In 1773, the irreverent *philosophe* Diderot went to St. Petersburg as the first librarian of Catherine's newly formed library. His boldness, eloquence, and atheism promised so to corrupt the younger members of the court that the older courtiers imposed upon the empress to still the tongue of the libelous Frenchman. Catherine reluctantly agreed, but wished it to be done without her. It was, therefore, arranged that Diderot should meet "a Russian philosopher, learned mathematician and distinguished member of the Russian Academy, who was prepared to prove to him the existence of God, algebraically, and before the whole court."[16] (A project, by the way, continued by the mathematical genius, Kurt Gödel.) The occasion arrived, the whole court was present, and the Russian philosopher (Euler!) advanced gravely toward Diderot, and spoke in an earnest voice, full of conviction, "Monsieur, $(a + b^n) / z = x$, therefore God exists: answer that!" Diderot, though unmatched in his own intellectual domain, was not Euler's equal in mathematics, and so withdrew from the contest. Shortly thereafter Diderot thought better of his position at the court of Catherine and returned to France. Even the encyclopedist Diderot joined the ranks of the German princess when confronted by the pronouncements of a modern mathematical physicist.

Insubstantial mathematics and materialist world conceptions that longed for mechanical models of unseen realities: these were the features of the mental world of the eighteenth century. Earlier light had inhabited other and very different imaginal landscapes, but now its nature had of necessity to take up residence in the psychological space offered to it by the best minds of the period. Is it ever otherwise? Does not our tenor of mind shape the world we see?

Yet the human spirit is restless and nature forever compliant, willing to answer as yet undreamed questions, capable of opening up vast new vistas, revealing still undisclosed parts of her being. Puzzling features, unexplained phenomena of light remained and encouraged continued investigation. Under their influence, a Frenchman devised a theory so successful that it overthrew Newton's image of light and established his own wave theory as the supreme imagination of light.

Light at the Roadside

Diffraction was a recalcitrant fact that pointed incessantly toward a wave theory. However, the puzzle of "polarized light" also resisted comprehension and seemed initially to favor a corpuscular view. And for advocates of the wave picture, above all else, what in fact was the all-pervading ether? These issues drove the research efforts of wave theorists to modify Euler's views such that the wave model of light became a formidable and finally triumphant adversary of the corpuscular view. Young's principle of superposition, as bizarre as it might have seemed to some, contained the essential missing ingredient. It still waited for a gifted, mathematically minded spirit to cast the principle into an elegant and powerful formalism. Ironically, that figure was to be found at the roadside supervising the design and construction of roads and bridges—Augustin Fresnel of the *Corps des Ponts et Chaussées*.

—

IF ONE NEEDED to point to a single event that marked the crossing to a new mathematical imagination of light, it would have to be the March 1819 sitting of the Paris Academy of the Sciences. Among the members of that committee were the greatest mathematical physicists of the century, most of whom were staunch corpuscularists. They were to judge the best scientific treatment of the intractable problem of diffraction. Only two submissions were received, one absurd and the other a treatment of such power, scope, and mathematical sophistication that its author, a relatively obscure provincial French engineer, and his theory burst into the forefront of research into the nature of light.[17]

Alone and unaware of the work of Thomas Young, the civil engineer Augustin Fresnel had busily pursued beautifully designed experiments on diffraction (using the local blacksmith as his fine instrument maker) and developed the mathematical foundations of a wave theory of light that could account for what he saw. All the great mathematical physicists of his day, such as Poisson, Biot, and Laplace, were supporters of a Newtonian conception of light and therefore strongly antagonistic to the wave theory of light. Fresnel had to submit his prize treatise to these men to be judged, and over the years it would be against them that the unknown engineer from the provinces leveled his attacks in one scientific paper after another. By his sophisticated use of the principle of interference, together with his masterful application of the calculus, Fresnel made new and marvelous predictions—many of which were confirmed. Yet his analysis was not always complete.

In his contest paper, Fresnel included formal solutions to diffraction problems so difficult that he could not solve them for individual cases. In other words, he only analyzed the diffraction problem abstractly but could make no concrete experimental predictions. His brilliant adversary Poisson proceeded to solve one of Fresnel's intractable equations and then tried to use it to show up an apparent absurdity in Fresnel's theory. Poisson pointed out that Fresnel's own theory clearly predicted a spot of light *directly behind* a small opaque obstacle. Shine light, for example, on a BB and directly behind it, in the middle of its shadow, there should be a spot of light as bright as if no BB were there! Absurd, declared Poisson. The experimentalist Arago, who was a friend and supporter of Fresnel, did the experiment and found a spot of just the kind predicted by Poisson using Fresnel's theory. Fresnel's wave theory was vindicated. Poisson inadvertently had driven home the last nail into the coffin that carried the corpuscular theory of Newton.

Where in common life do such diffraction effects show up? Although studied by Fresnel using special light sources and instruments, a completely analogous phenomenon can be seen many nights around the full moon! As gossamer clouds drift over the moon, close around it appear rings of color: bluish near the moon, becoming white further out, and ending in a reddish band. Called the aureole or corona, this lovely sight is caused by the diffraction of moonlight around water droplets or ice crystals in just the same way Poisson's spot is formed, but now repeated for each of the cloud's many millions of droplets.[18] Instead of only obscuring light, the droplets and crystals of the cloud can put light where none should be if one imagines light as geometrical or corpuscular rays only. The aureole should remind us of the impossible feat performed by light, its uncanny ability to appear in the heart of the darkest shadow.

Fresnel and Young managed to explain another stubborn bit of experimental data, namely the phenomena of polarization. The most common instance of polarization is associated with glare. Whether driving or boating, the reflection of sunlight from surfaces can be annoying and even hazardous. Such glare can be magically reduced by wearing so-called Polaroid sunglasses. The secret of this innovation relies on the understanding of polarization developed largely by Fresnel.

Pass light through a clear crystal of Icelandic spar and it seems to become in some sense "oriented." This feature can be evidenced by passing the oriented light (or polarized light) through an identical second crystal. It, too, is completely clear, and yet for certain relative orientations of the two crystals, *no* light passes through the second one. Corpuscular theorists had suggested that particles of light, in passing through the first crystal, were selected according to their shape. If the second crystal was aligned in the same way, then light passed through, otherwise not. It would be like trying to put square pegs into square holes; they only fit if oriented certain ways. Polarization phenomena stumped wave theorists because sound does not show any polarization effects. If light was a wave just like sound, then they should show identical effects. Fresnel suggested a solution. If the vibration in the ether, which he viewed as light, were "transverse" to the direction of propagation, then one could accommodate an orientation.

Sound is a wave of compression and rarefaction in air. Speak into one end of a pipe and the wave propagates through it at the speed of sound in air. The wave motion is *along* the direction of compression. If a domino chain could be equipped with springs to reset each domino as it fell, the wave front of falling and reviving dominoes would be a good image of sound.

Light is more like waves on a string or on water than sound waves. A string can vibrate up and down, or left and right. These correspond to two different linear polarizations of light vibrations and so account for the "orientation" of light.[19] The elimination of glare, which is largely light polarized horizontally, can be accomplished by orienting polarization filters to pass only vertically polarized light.

One of the most surprising and beautiful polarization phenomena is that of polarization colors. To see them at work, simply crush an old-fashioned piece of cellophane or a transparent candy wrapper between two Polaroid filters (if you have an old pair of Polaroid sunglasses, you can just pop out the lenses). A glorious set of colored swatches suddenly springs into view and their colors change as you rotate the Polaroid lenses! Close

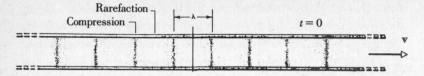

Depiction of a sound wave in air. Sound, moving to the right, is a series of compressions and rarefactions, as shown above.

—

observation shows that with every change in the thickness of the cellophane, a different color appears.

—

SEEING THESE COLORS, I cannot help being reminded of the "Synchromy" paintings of the American artist Morgan Russell. In his paintings, color itself becomes a language divorced from figure. Under one, called *Synchromy in Blue-Violet*, the catalog inscription reads, "Then God said, 'Let there be light!' and there was light. . . ." At the painting's center and dominating it is a yellow that Russell spoke of as symbolizing that light and the first eye that saw it. To Mrs. Whitney he wrote, the "bursting of the central spectrum in my picture . . . has surely a vague analogy with what must have happened to the first visual organ." In the teens of this century, Russell was also occupied with the construction of what he called "light boxes." These were small wooden boxes whose two longer sides were left open. Lamps could be mounted within the boxes, and painted transparent tissues were then mounted on the open sides. When lit from within, Russell's art literally glowed, radiating color into space like the rays of sunlight through panels of stained glass. Discontent within the confines of a box, light moves out and, where possible, through its surround, calling into life the otherwise mute, dark colors of tissue or glass.

On May 16, 1832, the great English astronomer William

Herschel sent to his physicist friend Whewell what he, too, called "a box full of light." It lacked the gaily painted sides and artistic pretensions of Russell's light boxes, but, Herschel explained, the carton was "not an *instantaneous light* box but one that works slowly." It contained the scientific papers of Augustin Fresnel.

Between 1820 and 1835, British scientists studied and applied Fresnel's ideas. As they did, it began to appear as if the essential mathematical relationships that characterize light had finally been captured and set down in Fresnel's papers, all of which could be contained in a modest-sized cardboard carton. Galileo would gladly have stolen into the dark box of Fresnel, perhaps with a candle, to discover in that solitary cell the mathematical truth about light. And yet for all the magisterial successes of Fresnel's theory, light was not at home or at peace in Herschel's box of light. It wanted out both scientifically and spiritually. The electrical experiments of Faraday and the dynamical theory of Maxwell tore open the box and scattered the papers. Is light a wave? Maxwell and Faraday would answer yes, but of what is it made, what kind of wave? And contemporary with Young and Fresnel were the thinkers, poets, and artists of Romanticism and American transcendentalism for whom the essence of light could never be caught in equations or put into a box. Their revision of light was far more angry and radical than that of the scientists, and it, too, will form an integral part of our history of light.

The Death of the Material Ether

Discontent not to see, we strain every instrument and intellectual muscle to make out the nature of an invisible, everyday thing— light. Not seeing it, we speculate. Perhaps, like sound, it is a fleeting figure whose form speeds through space on a quivering medium. As sound is carried on the air, and ocean waves on water, perhaps light is a wave borne by a supposed ether. Resisting direct observation, light was imagined by most physicists throughout the nineteenth century to be the vibration of an ethereal matter, but there were significant problems with such a view.

Air and water we know, but what is the character of this material ether? It seems hardly satisfactory to explain one invisible thing, light, by another, the ether. What are its density, texture, consistency, and other physical properties? From the fact that we and the earth are hurtling through space, and so through the ether, without apparent effect, it must be extremely subtle. Yet the enormous velocity of light, 186,000 miles per second, requires of the ether other, seemingly contradictory properties. Again we can proceed by analogy, this time with waves on a string or rope.

Stretch a long rope between two posts. Tap one end and you will notice that the disturbance moves swiftly to the opposite end, where it reflects back and returns to its origin, again to reflect and strike out again. . . . Draw the rope more taut and repeat the experiment. The form moves faster now. Slacken the rope and it moves slower. Clearly the speed with which the disturbance travels depends on the tension in the rope. We sense the rightness of this relationship from our own experience. Snap your fingers. If you snap them with more force, your fingers move more quickly. Similarly the rope's tension is that force which acts to restore the rope to its original position. The larger the restoring force, the quicker the rope snaps back into place, and the faster the disturbance scoots along.

Speed increases with tension. However, a second feature of the rope also affects speed, and that is its mass. It is harder to snap back a heavy rope than a light one under the same tension. Not surprisingly, therefore, as the mass goes up (actually the mass per unit length) the speed goes down. On a piano, guitar, or violin, instrument makers take advantage of this relationship using heavy strings for the bass notes and thin ones for the high notes. Careful experimentation and theoretical analysis show that the speed of the disturbance is given by the formula $v = \sqrt{(T/m)}$. The speed v is equal to the square root of the tension divided by the mass. Increase the tension and the disturbance speeds up, increase the mass and it slows down, in exact accord with the above equation.

By analogy we can ask of the ether, if light is a wave within this elusive medium, then it, too, must have a source of tension and a mass density. What are they? The answer puts flesh onto an otherwise vague notion, providing us with a concrete material image of the ether and so, too, of light. Everywhere light reaches, the ether exists also, sustaining the vibrations that are light. It is an elastic solid so resilient that a wave traveling in it can circle the earth over seven times in a single second. Yet that same earth must be capable of passing through the ether atmosphere of the universe unimpeded.

Already in 1746 Euler had made explicit estimates of the physical properties of the ether based on the comparison of the propagation velocities of sound and light. On these grounds he argued that the density of the ether must be at least a hundred million times less than that of air, and its elasticity a thousand times greater, to account for the extreme rapidity of light. Thus the ether was to have a resiliency greater than steel, but was also millions of times more subtle than air.

A century later, Sir George Stokes suggested a model of the ether at the close of a technical paper.[20] Glue and water together form a stiff jelly that can on the one hand act like a solid for *rapid* vibrations, yet will allow the easy passage of a *slowly* moving body. Perhaps the ether is of a similar substance. The exceedingly rapid, small vibrations of light race through the medium at 186,000 miles per second, while the ponderous motions of the planets creep along their orbital paths at a mere 10,000 miles per hour, plowing the ether aside as they pass.

Following Fresnel's lead, able French and English mathematicians created dynamical models of the ether like Stokes's, that is to say, they elaborated detailed pictures of the motions and interactions of "ether molecules" required by the known properties of light. From light's incredible speed, requirements had been placed on the ether's elasticity; from the motions of the planets and comets, limitations were put on the ether's density; from the facts of polarization, the structure of ether interactions could be suggested. Following on many experimental results, the work of these great mathematical scientists appeared to be converging on an elaborate material and mechanical model of the luminiferous ether that could account for every single experimental phenomenon.

Their picture of light possessed only one absolutely fundamental failing, namely that the ether was material. Most scientists of the nineteenth century were locked into a mode of imagination that was rigorously materialistic. No matter how subtle and unusual its properties, if the ether was some*thing*, then that thing must possess a substantial nature of some kind. If it did not, then it simply could not exist. For them the opposite of matter was spirit, and to admit the immateriality of light was to open the floodgates of speculative natural theology.

In the middle of the eighteenth century, Bishop Berkeley and Chevalier Ramsay had each advanced views of the ether as essentially spiritual, reaching back to the *prisca sapientia* (primal knowledge) tradition of Egypt, Greece, Persia, and the Hermetic writings already familiar to us from the previous chapter. Ramsay called the ether "the *body* of the Great Oromazes [i.e., Ahura Mazda], whose soul is truth. . . . He diffuses himself everywhere."[21] Trinitarians made the analogy between, or suggested the identification of, the Holy Ghost and the Universal Ether. After 1875, such sentiments received renewed consideration when a "crisis of faith" struck many late-Victorian scientists, who then often sought to reconcile science and religion by means of spiritualism.[22]

To deny a material nature to light or the ether was, therefore, to court a reversion to prescientific, spiritual imaginations of light, a view that the reigning scientific minds of the age abhorred. We should not underestimate the significance of religious inclinations (or disinclinations) in the conduct of science, especially with regard to the advocacy of poorly supported hypotheses. In this context we can understand the insistence of most scientists of the period that light and the ether must at root be material. The detailed dynamical models of the period offered them, literally, a concrete understanding of light that satisfied their metaphysical bias, and moreover could be used to make predictions as to how the ether would show itself, even if only in extremely subtle ways.

The ether was accordingly systematically searched for in numerous laboratories and observatories both in Europe and among the fledgling American scientific community. Every suggested experiment was done and redone, and even today new experiments in search of a material ether continue to be performed with magisterial precision. By 1900, the indications were growing clear; by 1990, they are undeniable. The material ether does not exist. It was a hypothetical fiction born of a materialistic imagination.

Light is *not* a luminous ripple on the material substrate of the ether. Still, although innumerable experiments deny the ether, an equal number seem to affirm the wavelike character of light. If we take both seriously and suppose light to be, in some sense, a wave, then what is it that is waving? In the cases of water waves, sound waves, vibrating strings ... some*thing* is always waving. The figure of sound is borne by the air. What bears the fleeting figure we call light? One thing has become certain, whatever it is, it is not material!

CHAPTER 6

Radiant Fields:
Seeing by the Light
of Electricity

For the invisible things of Him from the cre-
ation of the world are clearly seen, being
understood by the things that are made, even
his eternal power and Godhead.

St. Paul, Romans 1:20

Well before the bankruptcy of material imaginations of light
became evident, another and more productive view dawned in
unexpected quarters.[1] Far from the ancient and learned halls of
Oxford and Cambridge, and therefore also far from the pompous
tyranny of academic tradition and rivalry, a failing blacksmith
and his gentle wife had a son, the third of four children. The
year was 1791, the surroundings were poor and inauspicious—
a London slum. The ailing father eked out a spare existence as
a smith, his family living above the coach house. The daily fare
was modest. In 1801, when food prices were high, the young

son would be given a single loaf of bread to supply his meals for the week. Mean outer circumstances, however, can sometimes belie remarkable destinies.

From these unlikely beginnings, Michael Faraday, the greatest experimental scientist of all time, arose. By the time of his death at age seventy-six, he had been elected an honorary fellow to the scientific societies in Paris, Brussels, St. Petersburg, Florence, Copenhagen, Stockholm, Berlin, Munich, Vienna, etc., and been offered knighthood as well as the presidency of both the Royal Society and the Royal Institution of London, honors that he steadfastly declined. When his much-loved colleague pressed the aged Faraday to accept them, he replied, "Tyndall, I must remain plain Michael Faraday to the last."[2]

From this "plain" soul would come a revolutionary conception of light, one not fettered to the materialistic worldview of his time. Historians have traced to this simple son of a blacksmith the seminal imaginative ideas of modern field theory. I see in him an uncommon figure who cared too deeply about truth to be beguiled by the fashionable models of the day, and who knew that nature was the constant arbiter of ideas. How did it come about that plain Faraday saw so deeply into the heart of nature? What was his character, his education, and the special gift that made him the pivotal individual of the nineteenth century for our biography of light?

To judge from his childhood and adolescence, Faraday's successful career of scientific discovery seemed improbable in the extreme. He was a Cinderella figure rising from the ashes of his father's smithy to become a prince of natural philosophy. Michael Faraday's formal education was halted after he had learned the bare rudiments of writing, reading, and arithmetic. He was withdrawn from school by his tenderhearted mother after being beaten for mispronouncing the letter "R." Henceforth he grew up at home and on the streets of London.

Although the family was poor in things material, they led a

rich religious life. The latter proved to be an especially formative and lasting influence on Faraday. From those early years he maintained a quiet and unswerving religious faith as practiced within the small Christian sect into which he was born, the Sandemanian Church. Throughout his life, here was Faraday's spiritual home. Faraday saw no disagreement between his work and that of his God, for, paraphrasing a favorite Sandemanian passage from St. Paul, Faraday wrote, "even in earthly matters I believe that the invisible things of Him from the creation of the world are clearly seen."[3] The passions of objective scientific research and intimate religious sentiment alike moved through the breast of Michael Faraday. In subtle yet significant ways each played into the other, so that his agnostic colleague Tyndall could write of his mentor: "The contemplation of Nature, and his own relationship to her, produced in Faraday a kind of spiritual exaltation which makes itself manifest here. His religious feeling and his philosophy [science] could not be kept apart; there was an habitual overflow of one into the other."[4]

One of the "invisible things" Faraday investigated was light. Following St. Paul's precept, he sought knowledge of the invisible by understanding "the things that are made," that is, by constant, open-minded observation and experimentation. Nothing was so important to him as a truly revealing phenomenon. Faraday labored with unflagging energy and genius toward the archetypal phenomena of electromagnetism. Unknown to any before Faraday, once found, these phenomena suggested a totally new imagination of light.

Electric Waves

At the age of thirteen, Michael Faraday launched upon his career as an errand boy for Mr. G. Riebau of 2 Blandford Street, a French immigrant bookseller and bookbinder. Among the books of the shop and under the kind encouraging eye of Mr. Riebau, Faraday's curiosity flourished. Within a year Faraday had signed on as an apprentice bookbinder, a trade he mastered well over the next seven years. But his education at the shop went far beyond matters of bookbinding. While his hands received their training in dexterity, for which he was admired throughout his life, his spare moments were spent in reading. "There were plenty of books there, and I read them," wrote Faraday simply. Riebau also recollected how Faraday would copy drawings of electrical machines and the like into his own volumes, and how many mornings he would take an early walk, "Visiting always some Works of Art or searching for some Mineral or Vegetable curiosity—Holloway water Works, Highgate Archway..."

Faraday's energies received a more specific focus when, in 1810, at age nineteen, he joined the City Philosophical Society, which had been founded but two years earlier under the guiding spirit of Mr. John Tatum, whose house, library, scientific equipment, and abilities as a lecturer supported the Society. Every Wednesday evening Tatum or another club member would speak to a scientific theme that concerned him. Faraday's practice of writing out and binding a meticulous summary of the lectures, including drawings, in the end stood him in good stead.

Mr. Riebau occasionally showed Faraday's four bound quarto volumes of notes to friends and customers, including a Mr. Dance. On seeing them, the latter arranged for the young Faraday to gain admission to the lectures of Sir Humphrey Davy of the Royal Institution, the most famous scientist in England and a charismatic speaker. The effect of the lectures was to convert Faraday into a disciple. He proceeded to write up Davy's four lectures with the same care he had Tatum's.

Shortly thereafter, in the autumn of 1812, Faraday's apprenticeship with Monsieur Riebau was at an end, and before long Faraday was in despair. He longed to be a scientist but saw no way of fulfilling his aspirations. In desperation, he wrote to Sir Joseph Banks, president of the Royal Society, to apply for a scientific situation, no matter how menial. Banks did not even reply to Faraday's repeated inquiries. Yet fate worked in remarkable ways. Humphrey Davy suffered injury to his eyes through a laboratory explosion, and probably through the recommendation of Mr. Dance, Faraday came to serve as Davy's amanuensis for a few days. In December, Faraday wrote to beg a position of Davy, including with his entreaty his bound volume of meticulous notes of Davy's lectures. Though flattered, Davy (who was also of low birth) could offer him nothing. But again circumstances chanced to move in Faraday's favor. Davy's assistant became involved in a brawl and was dismissed. That very evening, as Faraday was undressing in his bedroom, a thundering knock startled him. A footman descended from a carriage with a note from Sir Humphrey Davy requesting Faraday to call on him at the Royal Institution the next morning. At the interview, Davy offered him the menial position of assistant, a guinea a week, two garret rooms atop the Institution, fuel, and candles. Faraday asked only that he be provided with aprons and, most important, be granted liberty to use the apparatus of the Institution. Davy and the Institution agreed and Faraday's apprenticeship in science with England's greatest chemist was begun March 1, 1813.

Faraday's dexterity and energy quickly showed, and his services were sought on many fronts in the research and lecturing activities of the Institution, where Faraday played the role of assistant to all. During his first three years, he accompanied Davy on his travels to European laboratories as his valet and assistant, meeting in the process the great scientists of France and Italy. It was not long before Faraday's own mettle began to show through a series of modest chemical publications, and then the discovery of benzene. But our specific concern with Faraday is his central role in providing a new imaginal form for the

One of the many beautiful Chladni plate patterns. Sand accumulates on those parts of the vibrating plate that are still.

nature of light, one that ultimately broke free from the materialistic constraints of his contemporaries.

Perhaps not coincidentally, the period of Faraday's interest in the nature of light occurred simultaneously with his research into the nature of sound, music, and musical instruments (1828–30), an interest that dated back to his days in Riebau's shop, where singing was a favorite pastime. Of special interest to Faraday were the so-called Chladni figures.

Some years earlier, in 1785, Chladni had discovered that beautiful patterns could be made to appear on thin metal plates when sand was sprinkled onto their surface, and then the plate set to sounding by bowing an edge like a stringed instrument. Faraday studied these lovely phenomena and demonstrated them to his audiences at the Royal Institution.

In just these years Fresnel's important articles on the undulatory theory of light appeared in English in a popular form whose mathematical level was sufficiently low that Faraday could read the articles with understanding and pleasure, admiring Fresnel's clarity and precision. In a similar vein, Herschel, in his 1830 treatise on natural philosophy, emphasized again and again the similarities between sound and light, making full use of the Chladni figures as an aid to the imagination. The analogous character of light and sound initially argued by Euler and others in the previous century was now very much in the air.

Faraday's religious convictions led him to believe deeply and steadfastly in the unity of nature, in the idea that what appeared on the surface to be disparate was at its core one. Perhaps vibration was such a unifying idea under which not only sound and light could be joined, but electrical effects also. To this end he energetically pursued a series of investigations in search of an undulating electrical wave of some kind. His discovery of just such an effect, one of the most important in his life, goes under the name of "electromagnetic induction."

As a name only, it is unfamiliar to most people outside physics and electrotechnology, but in its applied form it is universally familiar. Within almost every appliance and atop millions of telephone poles are transformers whose lineage goes back to that moment in August of 1831 when Faraday discovered the principle behind these devices. The impact of Faraday's discovery both practically and for our understanding of light is so great that we must pause to examine it. For in attempting to understand what appears to be a purely electrical effect, the foundations were inadvertently set for a new understanding of light. The effect appears in two important ways.

For his first experiment, Faraday wound two separate coils of wire around a torus (or doughnut) of iron. He connected one to a sensitive meter whose needle deflections would show when a small current was passing through the coil. To the second coil he connected a battery through a switch. With the switch closed,

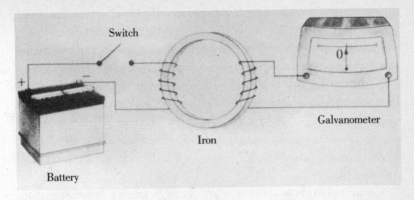

Faraday's archetypal experiment concerning electromagnetic induction. When the switch on the left is closed, a current surges through the left coil, which induces a current to flow in the circuit on the right, which deflects the galvanometer needle.

current flowed through the left-hand coil: with it open, there was no current. While watching the meter, Faraday opened and closed the switch. Neither with the switch open nor with the switch closed did the meter show a deflection. That is, there was no current in either case in the coil on the right. However, at just the moments of closure and opening, in that instant, there was a sharp deflection of the needle. Moreover, the needle deflected in opposite directions depending on whether the switch was being opened or closed. This showed that a brief surge of current appeared in the right coil when one *started* or *stopped* the current flow in the coil on the left. In other words, current was induced in the right coil only when there was a *changing* current in the coil on the left. Change in one circuit induced change in another. By contrast, when the current was unchanging, there was no effect on nearby coils.

One final caveat concerning the nature of the connection between coils. While the above "induction" phenomenon is enhanced by the iron torus core linking the two coils, the torus is by no means essential to the experiment. No electrical current passes through it. Remove the iron core and a small but still detectable induction effect is seen. A change in the electrical state in one coil induces a corresponding change in the other.

To understand electromagnetic induction, Faraday suggested that a *"wave of electricity"* is caused by sudden changes in the current through the first or "primary" circuit. This electrical wave travels through space and induces a similar disturbance in the nearby "secondary" coil of wire. Hence the deflection of the needle.

Clap your hands and the disturbance echoes through space, strike a match and darkness progressively gives way to flickering light. Likewise, throw a switch on a powerful electrical source and the surge of current that follows causes an invisible electrical disturbance to propagate through space that can be caught in a web of circuitry some distance away. When pushed to its full potential, Faraday's first experiment metamorphoses into the radio communication of a space probe sending its pictures back from Jupiter, Saturn, and Uranus hundreds of millions of miles away. Create an electrical disturbance in the on-board antenna of the spacecraft, and three hours later a tiny but exactly corresponding disturbance arises in a sensitive earth-based antenna. The investigations and ideas forged by this blacksmith's son continue to propagate new technologies and new knowledge.

Yet, just what is this "electrical wave" that it can connect distant circuits without a visible material connection of any kind? Michael Faraday's efforts to answer this question by further experimentation and cautious speculation would occupy him for nearly thirty years. The suggestion he finally put forward ultimately transformed our scientific image of light completely.

The second and related experiment, performed hot on the heels of the first, replaced one coil of wire (the one connected to the battery) by a magnet. Faraday discovered that by moving a magnet into and out of a coil of wire, a small current could be set to flow. If the magnet was stationary relative to the coil, no current; when it moved, current arose. This is one of the great archetypal phenomena of electromagnetism. Its practical and theoretical significance are hard to exaggerate. The production of electrical power at every generating station is nothing more than this action, in one form or another. Not only did industry make good use of Faraday's discovery, but so would pure science down to Albert Einstein, who used Faraday's archetype at the turn of the twentieth century when putting forward his "principle of relativity."

Natural Truth and the Shadow of Speculation

The view I am so bold as to put forth considers, therefore, radiation as a high species of vibration in the lines of force....

Michael Faraday

The worldview held by most early-nineteenth-century scientists was straightforward. The universe was filled with material objects, between which stretched several elusive but material ethers whose motions conveyed the forces of gravity, light, heat, electricity, and magnetism from one object to another. Everywhere there was substance: ponderable, massive matter. Never before or since has materialism attained a more contented and comprehensive grip on the world. The radiant eyes of the Egyptian god Ra that once lit a civilization were tightly shut.

Into this arena stepped the slight figure of Michael Faraday with views that would overturn the fundamental conceptions of his own beloved scientific community. Gentle, deferential, and religious man that he was, Faraday hardly cut the figure of a revolutionary. He was, however, a persistent experimenter and thinker. To resist the tide of current scientific opinion, which favored Fresnel's vibrational ether theory of light, would be heresy. Yet in two addresses to audiences at the Royal Institution, the first in 1844 and the second in 1846, Faraday dared to speak, and so carried science across a divide in the Western imagination of light.

From his earliest researches in science, Faraday possessed a love for truths gotten through direct experience, and consequently he held an abiding distrust of speculative theories, such as the molecular ether theories popular in his day. Ever aware of the danger of such speculation in his own research, Faraday penned the following entry in his diary for December 19, 1833, shortly after his discovery of electromagnetic induction: "I must keep my researches really *Experimental* and not let them deserve any where the character of *hypothetical imaginations*."[5] The key word here is "hypothetical." Time and again new ideas and imaginations must be grounded in honest experimental fact, otherwise fantasy takes the place of cautious creative thinking.

In his discourses of 1844 and 1846, Faraday very cautiously advanced his own views on the ultimate nature of matter, electricity, and light. His first attack was on the naive corpuscular view of atomic matter then current.[6] In place of atoms thought of as small "blobs" of impenetrable matter, Faraday suggested that atoms were purely "centers of forces." Since we know of objects by their attributes, and these are conveyed by forces only, we should not add the unnecessary concept of a material source to the forces.

Buddhist philosophers might put it this way. Can one think of an "attribute-bearer" apart from all attributes, such as size, shape, position, etc.? As an example, consider a penny. It is hard, round, thin, copper-colored, nearly an inch across, and carries embossed images on both sides. One by one, eliminate these attributes of the penny. Try it. First erase the images and the penny becomes a slug. Go on to imagine it as having no definite color, size, or shape. Can you do it? I cannot. If one cannot think of an object without its attributes, then why retain the notion of attribute-bearer at all? Similarly, if we only know of the world only through various forces, why posit the existence of force-bearers?

Since all properties such as hardness, color, and so on were understood to be the result of forces, then atoms (which Faraday did see as necessary) are simply the geometrical foci or centers of those forces. Substantial atoms, thought of as tiny, dense bits of matter, disappear completely, but as they do so, the atmosphere of forces that was thought to surround them becomes all-important. Force, not substance, is the true being of the world, and it, not the ether, reaches from one end of the universe to the other. Forces may constellate themselves in myriad ways and patterns to create chemical species, a visual, tactile, and fully sensual, "corporeal" world, but at root everything is still force, not ponderable substance. Faraday's ontology was very different from his colleagues'.

Recall Descartes's image of a cosmos filled with matter, a *plenum* that was like a whirling stream of water in which planets moved like floating bits of chaff, twigs, and leaves. Faraday suggested that Descartes's stream was actually a sea of pure forces. Points of matter—atoms—were only the starlike intersections of myriad raying lines of force that spread out from these centers to weave their way through the universe.

In his second address two years later, Faraday, apparently inadvertently, took his ideas an important step further.[7] According to the usual account, on April 10, 1846, Faraday and his collaborator Wheatstone were outside the Royal Institution's lecture hall waiting for the clock to signal the beginning of their lecture. It was Faraday's understanding that Wheatstone would be giving the lecture, but just moments before the evening's performance was to begin, the intensely shy Wheatstone bolted down the stairs. Caught unprepared, Faraday nevertheless jumped into the breach, speaking first on the scheduled topic of Wheatstone's "Electro-magnetic Chronoscope," but then following it impromptu with his famous "Thoughts on Ray-Vibrations."[8] Short of material on which to speak, Faraday "threw out as matter for speculation, the vague impressions of my mind."

His vague speculations were, while new to others, not new to him. They had been maturing during the previous fifteen years of his investigations. Already in the earliest of his reports on *Experimental Researches in Electricity* from November of 1831, Faraday had used the concept of "lines of magnetic force," the idea so central to his 1844 and 1846 discourses. If you sprinkle iron filings on and around a magnet, they align themselves in definite patterns. Faraday developed this picture into a powerful tool of the imagination. From his dogged investigations, and by careful reasoning, it seemed possible to him that one could conceive of the entire universe as threaded through with an infinity of such "lines of force." Reading his papers, one can watch the birth of a new scientific concept of the most extraordinary significance as it emerges from Faraday's experiments and imagination, that of the "field."

How is the earth bound to the sun? Perform the following thought experiment, Faraday suggests. Imagine a space with the sun alone. It resides solitary in its universe. Now, in an instant, place the earth ninety-three million miles away. Is it reasonable, he asks, to suppose that the distant presence of the earth causes there to rise up suddenly in the sun a power of attraction? Where would such a power come from? Would not such a conception violate our sense that forces must, in some sense, be conserved? Far better that the isolated sun, even before the earth's appearance, should have spread its influence as it does its light, even if no objects are there to feel it. Does the sun only shine when eyes are there to see? So, too, gravitational lines of force thread their way through space even before the earth appears.

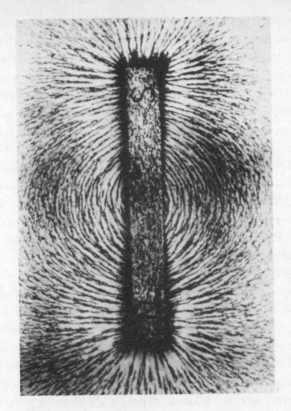

Visualization of magnetic field lines with iron filings.

The earth, then, reacts to a *local*, not a distant force, that is, to the force of the field at its own location.

In an era when all space was permeated by ethers, Faraday's suggestion could be interpreted to mean that in the ether were stresses and strains similar to those in the trusses of a bridge. The field idea, in this view, was but a superficial account of what was still, at root, a mechanical and material interaction. Faraday, however, chose to reach beyond the materialist imaginations of his contemporaries. Originally, he had considered the lines of force (his term for "fields") as nothing more than useful fictions, but the more he thought about them, the greater their reality appeared, until they became for him more real than the atoms of matter that were thought to be their sources, or the ether that supported them.[9] The results of his investigations into electromagnetic induction, moreover, convinced him of the existence of electric waves, that is, of the vibrations of such lines of force. He was, in fact, well prepared for Wheatstone's flight.

The revolutionary step taken in his 1846 lecture to Wheatstone's audience was his suggestion that the vibrations called light, so elegantly described by Fresnel and others, were not vibrations in an ether at all, but rather the vibrations of physical lines of force. Faraday's theory "endeavors to dismiss the aether, but not the vibrations."[10]

Vibrations were essential, the ether was not. Not content with transforming atoms into centers of force alone, Faraday also radically altered our conception of space and light. Lao-tzu's pot was emptied of its ethereal substances and filled with force, a force whose movements were light. Descartes would have stormed the Royal Institution if he could have gotten passage across Lethe and the English Channel, and even Faraday's most sympathetic reviewers thought he had, for once, gone too far. To make something as insubstantial as lines of force the ontological basis of the world seemed preposterous to the materialistic imagination, and yet herein lay the pregnant seeds of field theory as developed by physicists less than a century later. In the clear, if self-effacing, voice of the blacksmith's son, we hear the first sounds of a vast new vision of light.[11]

He knew his thoughts reached beyond the simple facts of the science he loved so much, and yet just those facts seemed to call out for an immaterial interpretation of light. Ever the gracious and careful philosopher, he advanced his view circumspectly, even apologetically, closing his remarks to those in attendance with the words "I think it likely that I have made many mistakes in the preceding pages, for even to myself, my ideas on this point appear only as the shadow of speculation. . . . He who labours in experimental inquiries knows how numerous these are, and how often their apparent fitness and beauty vanish before the progress and development of real natural truth."[12] His did not vanish.

Specters of Philosophy

But the splendour that supplies
Strength and vigour to the skies,
And the universe controls,
Shunneth dark and ruined souls.
He who once hath seen this light
Will not call the sunbeam bright.

Boethius,
from "The True Light"[13]

In the year A.D. 524, a lone philosopher-sage of Theodoric's court in Ravenna awaited his torture and death at the whim of his suspicious lord. He had used his years of study and meditation on the great Greek philosophers, as well as on the fruits of the new Christian faith, in service of the temperamental Theodoric. The solace of his studies now deserted him. Lost in self-pity, he was startled by a vision of a woman. Her stature was at once huge and common. Through tear-veiled eyes he saw her clear eye, quick sight, and grave countenance. Her gown, woven by her own hand, bore images of wisdom and the universe, as well as the signs of wounds made by those lusting for a snatch of knowledge. The divine Sophia, Philosophy, had come into the dark and solitary chamber to console a condemned man, the philosopher Boethius.

Inquiring, she discovered the real cause of Boethius' grief, his forgetfulness of his true self. Her balm, then, was to "dissolve this cloud with gentle fomentations that thou mayest behold the splendour of true light." In these, his final days, Boethius wrote his greatest work, *The Consolation of Philosophy*, so influential on the centuries to follow. When the book was completed, Boethius was hanged by a rope around his forehead and beaten to death with clubs. The last utterance of Philosophy in the *Consolation* reads: "Great is the necessity of righteousness laid upon you if ye will not hide it from yourselves, seeing that all your actions are done before the eyes of a Judge who seeth all things." These words were prophetic, for legend has it that after the execution of Boethius and his supporter Symmachus, a great fish's head was served to Theodoric. In his guilt-ridden mind, it took on the features of those whose death he had caused. He wept and anguished, inconsolable until his own death shortly thereafter.

Six hundred years later, the Chartres master Alan of Lille, in a period of mental anguish, was similarly visited.[14] The comforter called herself Natura, the being of Nature. Her every feature and attribute reflected the universe, her every word gauged to reawaken in Alan the knowledge of her that he and all the world seemed to have lost. Alan called the song she sang the *Plaint of Nature*, a song of violations unheard by humankind.

Six centuries later, still at the hands of Goethe, Faust in his high-vaulted Gothic study was lost in despair.[15] Having mastered every discipline through a long life of study, he felt himself a fool no wiser than before. In angry desolation, he called out to unseen forces: "Spirits that hover near to me, give me an answer if you hear my voice!" His eye fell on the signs of the Macrocosm and Earth Spirit.

Faust then invoked, even commanded, the Earth Spirit to appear. Unbidden, Natura had comforted and taught Alan of Lille, and Sophia before her had consoled the condemned Boethius. Faust, by contrast, bent the Spirit of the Earth to his

Faust in His Study by Rembrandt, 1652.

own needs and will. "Obey! Obey," he cried, "although my life should be the price!" When the Earth Spirit did appear, Faust reeled before its awesome shape of flame, but even so dared to declare himself a spirit kindred to this mightiest spirit of nature. Faust's arrogance was bluntly rejected by the Earth Spirit with the words, "Your peer is the spirit you can comprehend; mine you are not!" The spirit Faust could comprehend then entered his chambers; the plodding pedant Wagner, dressed in night-

gown, nightcap, and holding a flickering candle, a sorry counterpoise to the flame-formed Earth Spirit of a moment before.

The Earth Spirit seems to have grown ever more remote, more
difficult of access in these later days. The only way of approaching her, that of scholarship à la Wagner, most often leads
not to her but only to a shadowy, even demonic reflection of her
true self. Michael Faraday's great successor, James Clerk Maxwell, fought a Faustian battle as a young man, and confronted
a fearful counterimage of divine Sophia. The son of a distinguished family in Edinburgh, Scotland, and a very promising
young mathematician, Maxwell reported his own vision in Gothic
chambers when a third-year student at Trinity College, Cambridge, in 1852.[16]

The midnight bells had struck and Maxwell left his "confounded hydrostatics" to prepare for bed, asking himself doubt-
fully, "with voice unsteady, If of all the stuff I read, I / Ever
made the slightest use." He looks at his future, a life likely to
be full of outer accomplishment and lofty station, and knows it
will be but the fruit of reason put under "the control of worldly
pride." Across his papers, so littered with mathematical problems, spectral shapes flicker, becoming little marching creatures, the "glorious ranks" of professors past and present, "who
scrutinised the trembling lines." As if this were not enough, an
awful form then arose before the despondent undergraduate, a
creature intent on defending the timeworn practices of academia.

> *Angular in form and feature,*
> *Unlike any earthly creature,*
> *She had properties to meet your*
> *   Eye whatever you might view.*
> *Hair of pen and skin of paper;*
> *Breath, not breath but chemical vapour;*
> *Dress,—such dress as College Draper*
> *   Fashions with precision due.*

Eyes of glass, with optic axes
Twisting rays of light as flax is
Twisted, while the Parallax is
   Made to show the real size.
Primary and secondary
Focal lines in planes contrary,
Sum up all that's known to vary
   In those dull, unmeaning eyes.

As this wretched mathematical "Hag" proceeded to exorcise all feeling for beauty and poetry from the heart of Maxwell by demonic prayer and invocation, "Suddenly, my head inclining / I beheld a light form shining; / And the withered beldam, whining, / Saw the same and slunk away." In place of the hideous shape and voice of the "artificial spectre," he caught glimpses of "the being whom she aped." Maxwell does not call her by name but characterizes her as Boethius did his goddess Philosophy, and Alan of Lille his being Natura. Her modest, tranquil radiance far outshone the ill-fitting guise of pedantry, for no matter how pompously dressed, pedantry always veils, never reveals.

Yet, Maxwell declared, creation can withstand both reason and calculation, as long as one accompanies and adds to them the action of worship.

Worship? Yes, what worship better
Than when free'd from every fetter
That the uninforming letter
   Rivets on the tortured mind,
Man, with silent admiration
Sees the glories of Creation,
And, in holy contemplation,
   Leaves the learned crowd behind!

The young Maxwell longed to leave the learned crowd of Cambridge pedants behind, just as Faust, for all his learning, longed to leave Wagner to his dusty volumes in order that he himself might find truth. Such proved more difficult than the young Maxwell imagined.

Just a half year after his "vision," Maxwell fell seriously ill, and landed at the home of the Reverend C. B. Tayler, rector of Otley, near Ipswich, suffering from a "brain fever." During this time, Maxwell studied Sir Thomas Browne's classic *Religio Medici* (Religion of a Physician), in which the devout Anglican tread his way judiciously between heresies to a philosophic faith acceptable to the Church of England. Maxwell's ceaseless struggle to reconcile his genuine religious feelings with the rigorous demands of natural philosophy was sincere. The closing of his "Student's Evening Hymn" dating from this period reflects his pious sentiments.

> *Teach me so Thy works to read*
> *   That my faith,—new strength accruing,—*
> *May from world to world proceed,*
> *   Wisdom's fruitful search pursuing;*
> *   Till, thy truth my mind imbuing,*
> *I proclaim the Eternal Creed,*
> *   Oft the glorious theme renewing*
> *God our Lord is God indeed.*

In these lines we recognize Browne's two books of divinity: "Thus there are two bookes from whence I collect my Divinity; besides that written one of God, another of his servant Nature, that universall and publik Manuscript, that lies expans'd unto the eyes of all; those that never saw him in the one, have discovered him in the other."[17]

Although his youthful anguish over the schism between God and nature must surely have echoed in later years, the gap between faith and science became clearer and greater. Maxwell quickly came to hold that any attempt to harmonize the two "ought not to be regarded as having any significance except to the man himself."[18] It was a safe and conventional position given sanction by Luther and Calvin, and in our own day by the neoorthodoxy of the great German Protestant theologian Karl Barth.

The divine being of nature, as also the being of light, faded from view in the centuries and millennia that separated the luminous Egyptian eye from the ether whose vibrations comported themselves to the differential equations that Maxwell was destined to discover. More often than not, Natura gave way to the mathematical Hag. The centuries separating Alan of Lille from Maxwell marked the twilight of the gods. They, the gods, saw fit to remove themselves to regions less hostile than ours. As the German romantic poet Hölderlin put it: "Of course the gods still live, but right up there in another world!"[19]

Having segregated his spiritual from his physical interests, Maxwell was ready to mathematize Faraday.

The Electromagnetic Universe

The greatest alteration in the axiomatic basis of physics—in our conception of the structure of reality—since the foundation of theoretical physics by Newton, originated in the researches of Faraday and Maxwell on electromagnetic phenomena.

Albert Einstein[20]

Einstein's words prepare us for what is about to happen. The understanding of reality held by science, and increasingly by everyone since Newton, was changing profoundly. We are approaching the stunning culmination of that revolution with Maxwell's mature treatment of light. Until now, a mechanical universe of matter in motion had captivated the scientific imagination, but the researches of Faraday and Maxwell would ultimately entail a transformation of that worldview down to its very foundations. New powers, electric and magnetic, had been explored in the previous hundred years, and now our language, imagery, and mathematical sophistication were ready for their greatest achievement, the electromagnetic imagination of light. Maxwell accomplished the revolution without philosophical fanfare, but through the instrument of mathematical physics.

—

STATIONARY BICYCLES fall over. For that reason, we provide children with training wheels when they are first learning to ride one. They are not essential for learning the skill, and they make fast maneuvering awkward, but they do mollify fears. The ether was like a pair of training wheels for nineteenth-century physicists. It seemed obvious to everyone that wave motion was carried by some kind of medium, so one was cooked up. Maxwell was as sure about it as the rest. In his 1878 article on the "Ether" for the *Encyclopedia Britannica*, Maxwell wrote: "Whatever difficulties we may have in forming a consistent idea of the constitution of the aether, there can be no doubt that the interplanetary and interstellar spaces are occupied by a material substance or body . . ." otherwise how could light travel through those regions? Within thirty years, however, the ether would be abandoned. Once the electromagnetic theory of light got fully under way, the training wheels came off, and the new machine rode beautifully without them.

The most significant development to this end was Maxwell's mathematization of Faraday's scientific imagination. Time and again Maxwell spoke of his youthful resolve to understand all the known phenomena of electricity and magnetism first by reading all of Faraday's *Experimental Researches*, and only subsequently permitting himself to read the prevailing theory of his day. In Faraday's language, Maxwell wished first to read the book of "natural truth" before studying "shadowy speculations." Reading Faraday, Maxwell the mathematician was surprised to find a kindred soul, someone who thought mathematically even though he expressed himself in pictures. Faraday's invention and the use he made "of his idea of lines of force in coordinating the phenomena of magneto-electric induction shews him to have been in reality a mathematician of a very high order," wrote Maxwell.

Upon finishing his first early reading of Faraday, Maxwell resolved to translate Faraday's thinking into the language of mathematics. In this project he was singularly successful. His paper "A Dynamical Theory of the Electromagnetic Field," finished in 1864, is a landmark in the history of science. In it, Maxwell synthesized all of the disparate knowledge of electricity and magnetism into a single set of four equations, now called Maxwell's equations. Every electrical, magnetic, or optical experiment that has been done, or will be done (barring certain quantum effects), finds its formal theoretical explanation in terms of these four equations. Through his "translation" Maxwell demonstrated the extraordinary power and mathematical elegance of Faraday's ideas. About Maxwell's achievement, the Nobel laureate Richard Feynman once said, "Ten thousand years from now there can be little doubt that the most significant event of the nineteenth century will be judged as Maxwell's discovery of the laws of electrodynamics. The American Civil War will pale into provincial insignificance in comparison."[21]

For our purposes, however, Maxwell's great synthesis of electric and magnetic knowledge was important because it carried along with it an entirely unexpected stowaway. From the earliest beginnings of serious investigation into electricity and magnetism until the time of Maxwell, the studies of light and electromagnetism were entirely separate. Certain materials, like amber, attracted bits of paper after being rubbed. This principle was already known to the ancient Greeks, who called it the *elektron* effect, whence the name for the electron was derived. Benjamin Franklin had demonstrated that lightning was an electrical effect, and many were drawing lightning down into laboratories and even social parlors, playing with the newly discovered phenomena of electricity (a life-threatening pursuit, as some unfortunates found out). Countless amateurs and professionals alike were busy with the fascinating study of electrical effects. Other investigators were grinding lenses for telescopes and microscopes, and studying the intricate phenomena of light and color. What could these two realms of nature have to do with one another? Everything! was Maxwell's (and Faraday's) answer. This unifying vision more than anything else is what caused Einstein to call the researches of Faraday and Maxwell the origin of "the greatest alteration in our conception of the structure of reality since Newton."

In reading his 1864 paper, we can witness the moment of epiphany when the new structure of reality dawned on Maxwell. In it Maxwell develops an imagination in which every body is bathed not by a material ether, but by the electromagnetic field. Throughout space, surrounding and even penetrating all objects, are Faraday's lines of force. All electrical and magnetic effects (attraction, repulsion, induction, etc.) can be elegantly and exactly explained by Maxwell's theory within this view. Although brilliant, the synthesis of all electrical and magnetic effects is not the revolution of which Einstein spoke, but then Maxwell suggests the totally unexpected.

Toward the end of his paper, Maxwell turned his attentions away from electrical and magnetic effects to light, away from amber and magnetite, and toward the candle. The complex equations that he manipulated so skillfully, equations that completely baffled the older Faraday, suggested a bold hypothesis. For in his analysis Maxwell derived an equation for the electromagnetic field that was exactly analogous to the equation Euler had derived for the propagation of sound waves. Moreover, from that equation a prediction for the velocity of light could be made, and it agreed well with the best measurements then available.

In a moment of classic understatement, Maxwell concluded, "The agreement of the results seems to shew that light and magnetism are affections of the same substance, and that *light is an electromagnetic disturbance propagating through the field according to electromagnetic laws*." In this single sentence, Maxwell proposed a profound change in our image of light, one in which light, electricity, and magnetism would now, and forever after, be entwined. Two arenas of physics, which to all outward appearances have nothing in common, were to be united. Faraday's ray vibrations had come to maturity in the mathematical imagination of Maxwell. Light was declared an electromagnetic wave whose vibrations rippled through space.

Notice that Maxwell's thinking retained the ideas of the past, when he wrote that electricity, magnetism, and light were "affections of the same substance." That substance was the ether. Maxwell and everyone else still thought in terms of the ether, but his mathematical analysis did not require it at all. For the first time, the wave conception of light was given a full theoretical treatment that was, in fact, free of the ether. It would take four decades before the scientific community would fully realize this, but the ether was dead. Like a pair of training wheels, the ether had served a useful purpose, but could now be removed. The electromagnetic view of light rode all the better without them.

One who urged their removal was Heinrich Hertz. In 1887, he had experimentally verified that the laws of reflection and refraction, known so long for light, were also obeyed by the invisible electrical disturbances produced by sparks. Yes, light seemed to be but a special kind of electromagnetic disturbance. Hertz admired the mathematical formulation of light given by Maxwell, and warned against mistaking the "gay garments [i.e., the ether] we have used to clothe it" for the "homely figure" the real theory presents.

As the field concept gained acceptance Maxwell and others came to see the electromagnetic field itself as a repository of energy. Maxwell was totally clear about his conception of the energy associated with the electromagnetic field. "I want to be understood literally. . . . On our theory the energy resides in the electromagnetic field, in the space surrounding the electrified and magnetic bodies, as well as in those bodies themselves." The field possessed energy, according to Maxwell's theory. From a theorem due to Poynting, one was able to follow the flow of electromagnetic field energy explicitly. The results are very surprising. Why does a light-bulb filament get hot and glow? Not from the flow of electricity through the filament, but rather because as current flows, field energy streams into the filament from the space surrounding it. About this the mathematics seemed clear. Energy flows around and into conductors, not through them.

Could it be that at root everything is a form of field energy organized into various forms? Such was the question of Maxwell and other "energeticists." Impressive efforts were made in this direction, but floundered.[22] Only since Einstein and the development of quantum theory have we learned in what sense energy and matter are related.

—

FARADAY HAD BELIEVED in the eloquence of phenomena. He was a master inquirer, a rapt observer and ingenious experimenter who trusted what he saw. Besides the language of phenomena, two others were spoken: one imagination of nature was mathematical and the other mechanical. Faraday could not understand the former and distrusted the latter. Faraday entreated Maxwell to translate his treatment "out of their hieroglyphics, that I also might work upon them by experiment." His request was not answered. Faraday's distrust of mechanical models in favor of phenomena has also fared poorly in the intervening years.

The figure who, perhaps more than any other, labored longest and hardest to combine a mechanical picture of the world with the rigors of mathematics was William Thomson, later Lord Kelvin. With his efforts, the material ether and its associated imagination of light finally exhausted itself. In its wake, the natural truths of experiment encouraged fresh spirits to unheard-of conceptions of the being of light that reached beyond even those of Faraday and Maxwell.

The Final Failure

One word characterizes the most strenuous efforts . . . that I have made perseveringly during fifty-five years: that word is FAILURE.

William Thomson

In 1841, at the tender age of eighteen, William Thomson, already an enthusiastic reader of Faraday, entered the scientific community by publishing a sophisticated mathematical comparison of Fourier's theory of heat and electrical action at a distance. It was a paper that Faraday, thirty years his senior, could not have read with much comprehension.[23] Over the next decade Thomson's theoretical studies of heat established his reputation, and also convinced him that *all* physical phenomena, including light and electromagnetic forces, would be understood in terms of matter moving according to mechanical laws, if only one could discover the nature of the ether. It was a conviction held throughout his career, one he labored mightily to realize.

If he was right, then substance must be everywhere. "We now look on space as full," wrote Thomson, full of a fluid whose sole attributes he considered to be extension, incompressibility, and inertia. All colors, heat, electromagnetic effects, etc., are but the reflection of the more fundamental, lawful motions of this fluid. Based on the success of Fresnel's wave theory of light,

Thomson declared of the ether, "its existence is a fact that cannot be questioned," and went on to calculate its probable density from recent data on solar energy. Faraday's lines of force were wonderful, yes, but one needed to understand them in terms of the mechanical properties of the ether if one was to have any understanding of electric, magnetic, or gravitational forces. Thomson, therefore, launched a grand attempt to create a universal physics unified around the principles of simple, inertial matter moving in accordance with mechanical laws.

Light, as usual, proved the most recalcitrant feature of the world for Thomson. Maxwell, although sympathetic to Thomson's program, had been forced to abandon, at least temporarily, a theory involving the ether. Instead, he had retreated to a mathematical treatment for which there was no readily visualizable model. Thomson, by contrast, refused to give in. Adopting and transforming the conceptions of brilliant mathematical colleagues such as Stokes and Green, he fought his whole life long to find tenable mechanical hypotheses for the ether.

If you travel to Glasgow University, where Thomson lectured for over five decades, you will find there a demonstration begun by Thomson a century ago and still running. Its intention is to elucidate his view of the ether. Midway in a container of water, a slab of wax is lodged. Bullets are placed above the wax and small corks are located below it. After a year or so, the bullets slowly make their way down through the wax, while the buoyant corks migrate upward through it.

All of space was likewise to be understood as filled by a highly elastic substance that can vibrate at the exceedingly high frequencies of light required by Fresnel's theory, and which can, at the same time, allow entire planets and stars to make their slow and ponderous way about the universe. Stimulated by the work of his German colleague Helmholtz on fluid motions, Thomson proposed that space was filled with "vortex atoms" that move like smoke-ring structures in the ether. It was, in fact, a brilliantly conceived hypothesis, able to account for reflection, refraction, dispersion, and polarization phenomena. Yet, as the American physicist Willard Gibbs wrote at the time, for all its successes, and the audacious genius of its author, Thomson's theory "should not blind us to the actual state of the question. It may still be said for Maxwell's electrical theory, that it is not obliged to invent hypotheses but only to apply the laws furnished by the science of electricity."[24] Thomson had constructed but another idol to account for light, one no truer than those that had gone before. His hypothetical vortex atoms were short-lived and of little long-term value for the development of the science of light. Better, Gibbs maintained, not to invent such images but rather to stick to the pure abstract mathematical theory of Maxwell. Like the Muslims with their prohibition of images of God or the Prophet, Gibbs was concerned that the images of vortex atoms blind the scientist to the homely mathematical truths of light.

Gibbs was an ocean away from Thomson, a safe distance. For Thomson, now Lord Kelvin, lobbied hard for his views, and his powers of intimidation were legendary. Over the years, he had become the reigning patriarch of British physics. Even as an eighty-year-old man, Lord Kelvin remained a formidable presence, as is seen in a story told by Ernest Rutherford (discoverer of the nuclear atom) of his Royal Institution lecture on radium.[25]

I came into the room, which was half dark, and presently spotted Lord Kelvin in the audience and realized that I was in for trouble at the last part of my speech dealing with the age of the earth, where my views conflicted with his. To my relief, Kelvin fell fast asleep, but as I came to the important point, I saw the old bird sit up, open an eye and cock a baleful glance at me! Then a sudden inspiration came, and I said "Lord Kelvin had limited the age of the earth, *provided no new source [of heat] was discovered*. That prophetic utterance refers to what we are now considering tonight, radium!" Behold! the old boy beamed upon me.

Rutherford realized that the new heat source for the earth's remarkable warmth, even after millions of years, was the energy given off by the decay of radium. For eighteen years, Kelvin had fought geologists by "proving" from the principles of thermodynamics that the earth was young. He had, of course, understandably neglected to account for the undiscovered heat provided by radioactive decay in the earth's core, but one still had to be circumspect to avoid Kelvin's wrath.

For all his power, many of Kelvin's most passionately held convictions were to die in the face of simple experimental facts, the "natural truths" so loved by Faraday. One such fact that he lived to see was the careful experimental search for the ether performed by Michelson and Morley in 1887. The experiment showed unambiguously that there was no ether. From his own analysis, Kelvin could see nothing wrong with the experiment, and in his famous "Two Clouds" address of 1900, he called this experiment a very dense cloud hanging over the ether theory of light he favored. Only five years later the cloud was removed by Einstein's special theory of relativity, but with it there also disappeared the ether theory of light. The second cloud Kelvin saw concerned the spectral colors of light associated with incandescence. The understanding of heat (thermodynamics) and electromagnetism before 1900 was inadequate to account for the colors given off when so-called black-bodies were heated. This discrepancy would only be removed by the advent of quantum mechanics.

Lord Kelvin was prescient. From these two clouds over nineteenth-century physics would arise the twentieth century's extraordinary transformation of physics and a new image of light. A third cloud, however, also existed, one that would make its own significant contribution to modern physics.

Part of the solar spectrum.

—

RECALL THAT IN 1612 Galileo had pointed his telescope at the sun and seen dark blemishes passing over its surface, to the disbelief of contemporary astronomers. The self-educated, Bavarian instrument maker Joseph Fraunhofer (1787–1826) took Galileo's spots on the sun one step further. In 1814 Fraunhofer attached a telescope to a prism and examined the spectral colors of sunlight more carefully than Newton or anyone had done. The perfect continuity of colors seen by Newton, one blending imperceptibly into the other, was in fact broken by black lines. Whereas Galileo had seen dark blemishes on the bright surface of the sun, Fraunhofer discovered dark blemishes in the otherwise glorious phenomenon of the spectrum. What were these dark lines that marred the rainbow of sunlight?

Not until 1859 did two Heidelberg professors, Gustav Kirchhoff and R. W. Bunsen, establish the unambiguous connection between the patterns of dark spectral lines seen in sunlight, and the similar bright sequences of spectral analysis of light produced when vaporized materials were put into the flame of a Bunsen burner. Every element possesses a unique light signature, a unique spectrum of discrete lines.

With this discovery astronomers could for the first time turn their telescopes to the sun and stars, and, from the spectra they saw, determine the kinds and amounts of elements present at those distant locations. Terrestrial physics had, through the spectral analysis of light, made a leap into the cosmos. The sun and stars were seen to possess spectra just like those that could be produced in Heidelberg, leading to the simple conclusion that the stars differed little (aside from average temperature) from our familiar world.

The spectral analysis of light must join the Michelson-Morley and "black-body" experiments as a third cloud on Kelvin's otherwise triumphant, clear view of the horizon of physics. Nineteenth-century physics had succeeded brilliantly in nearly every domain. Yet unexplained phenomena of light persisted. It took courage to look at them carefully, but those who dared ultimately were rewarded beyond their greatest expectations.

—

IN THE FACE of these issues Lord Kelvin could, for all his bravado, be candid about his own accomplishments. During the banquet celebrating the fiftieth jubilee of his professorship at Glasgow, he responded to a toast of high praise with these words: "One word characterizes the most strenuous efforts for the advancement of science that I have made perseveringly during fifty-five years: that word is FAILURE. I know no more of electric and magnetic force, of the relation between ether, electricity and ponderable matter than I knew and tried to teach to my students of natural philosophy fifty years ago in my first session as Professor."[26]

Kelvin had steadfastly resisted the dawning of a new imagination of light. If one insisted on understanding light as material, then one was certain of failure. Kelvin, and those who thought like him, were forced to step down. The threads first spun by Faraday and woven by Maxwell needed still to be picked up by fresh minds, and fashioned in ways less fettered to conceptions of the past. Led by experiment and the prompting of genius, a new century of light dawned in 1900, one whose implications we are still struggling to comprehend.

We stand poised to cross the threshold of the twentieth century, yet I would have us draw back for a moment to consider more closely what has become of light, and the phenomena it presents to us.

Disturbances of the Heart

When the lamp is shattered
The light in the dust lies dead—
When the cloud is scattered
The rainbow's glory is shed.
When the lute is broken,
Sweet tones are remembered not;
When the lips have spoken
Loved accents are soon forgot.
                    Percy Bysshe Shelley

In the last years of his life, Faraday's mental powers gradually slipped away, sending him into a gentle senility. When his lone disciple, Tyndall, who self-consciously longed to play Schiller to Faraday's Goethe, made his last visit to the old natural philosopher (Faraday detested the new designation of "scientist"), he found that "the deep radiance, which in his time of strength flashed with such extraordinary power from his countenance, had subdued to a calm and kindly light." Tyndall remembered how he had "knelt beside him on the carpet and placed my hand upon his knee; he stroked it affectionately, smiled, and murmured in a low soft voice."[27]

Faraday's last speechless pleasures were the sunsets he could see from the gardens of his retirement house. In earlier years the sight of a rainbow, a storm, or a sunset would set his eyes aglow. Here were "natural truths" on a grand scale, truths worthy of our wonder and awe.

Yet throughout the nineteenth century, many asked, how did such phenomena look through the lens of the electromagnetic theory: Was the rainbow but a web of electromagnetic equations, the sunset but the effect of differential light scattering, was lightning only countless millions of electrons? Had Maxwell's mathematical Hag in the end gained the upper hand? Was it enough to be a devout Anglican while she pillaged the temples of nature, smashing gods and goddesses, only to raise up in their place the idols of the scientific imagination?

John Stuart Mill maintained staunchly that the love of natural beauty was no obstacle to scientific knowledge: "the intensest feeling of the beauty of a cloud lighted by the setting sun, is no hindrance to my knowing that the cloud is vapour of water, subject to all the laws of vapours in a state of suspension."[28] But is the inverse also true? Does not theoretical, scientific knowledge obscure and even slay the sentiments we have on seeing the sublime? Concerning another sunset, we recall Henry David Thoreau's words for Christmas 1851: "I, standing twenty miles off, see a crimson cloud in the horizon. You tell me it is a mass of vapor which absorbs all other rays and reflects the red, but that is nothing to the purpose. . . . What sort of science is that which enriches the understanding, but robs the imagination? If we knew all things thus mechanically merely, should we know anything really?"[29]

To Thoreau the brilliance of worldly light seemed to threaten the very existence of the spiritual light of imagination. Halfway around the world the Japanese philosopher Keiji Nishitani turned to the fourteenth-century Buddhist priest Musō Kokushi, who expresses himself similarly in his book *Muchū mondō* (Questions and Answers in a Dream). Musō recalls the ancients who maintain that every sentient being possesses a spiritual light drawn from "the samādhi of the Storehouse of the Great Light."

The "miracle-light" of all the Buddhas is drawn from this same source. In our own lives, every act of insight, even in such ordinary matters as distinguishing east from west, black from white, "is the marvelous work of that spiritual light. But fools forget this original light and turn to the outside in search of a worldly light."[30]

During Faraday's lifetime, the struggle was alive between those content with worldly light and others who remembered, even if only dimly, an original light. Goethe and Novalis in Germany, Coleridge in England, and Emerson and Thoreau in America; these poetic souls, far more than contemporary scientific minds, sought to understand science differently. Some romantic thinkers saw no way out but the rejection of science outright. Far more interesting were those others who followed Goethe's example and reexamined the essential nature of science to determine if there was really no room in it for man.

The "unweaving of the rainbow," as Keats would call it, had taken many centuries. Jahve had placed the first rainbow in the heavens when the sun dawned on Noah after the flood. As long as the rainbow stood in the sky, the covenant between God and mankind stood firm: "I set my bow in the cloud, and it shall be for a token of a covenant between me and the earth . . . that the water shall no more become a flood to destroy all flesh."[31] The rainbow was a promise of protection. Over the centuries, every optical discovery, and new conception of light, found application to the rainbow. To many poetic hearts, this was the "unweaving" of the rainbow at the hands of the mathematical Hag, an act that threatened to remove Iris from the heavens, and so to void the covenant between God and humankind.

Instead of merely mourning Iris's loss, could one not understand science in a way that left nature her soul, and human life its meaning? Together with the scientific breakthroughs of the early twentieth century—quantum mechanics and relativity—we will, therefore, listen to those who sought to reconceive science and light along more spiritual lines, and in a way that welcomed the being Natura even if she appeared in new dress.

First, however, we can draw together many centuries of light's biography by considering one of light's most beautiful effects: the rainbow.

CHAPTER 7

Door of the Rainbow

Then as we walked, there was a
heaped up cloud ahead that changed
into a tepee, and a rainbow was the
open door of it.

Black Elk[1]

In the spring of 1988 I saw an ancient rainbow. I had seen modern rainbows often in life, but this was my first experience of a rainbow that stretched back to the beginning of things.

In an annual rite whose origins are beyond recorded memory, the citizens of Gubbio, Italy, race to the monastery of St. Ubald atop the mountain on whose flank they live. In the early morning, three enormous, twenty-foot, phallic "candles," or *ceri*, crafted of wood, are shouldered by squadrons of men and boys and ceremoniously marched through the various quarters of the city. Atop each candle stands the flamelike form of a revered saint, sanctifying this pagan ritual.

From morning to afternoon, the preparations go on with feasting and merriment. Then with the sun low in the western sky, the *ceri* are taken to the town gate whose road leads to St. Ubald's basilica atop the summit. From my hilltop position in front of the church, I could see the *ceri* and, all along the winding road, the thousands of spectators and uniformed racers, the latter ready to shoulder the huge "baton" when their leg of the primal relay race would come. With a roar it began, and with amazing swiftness the tall, erect *ceri* wound their way uphill in a bizarre procession of grace and speed.

As they rounded the final curve all near me surged and yelled. The *ceri* passed by and we turned to watch them speed through the open church doors of St. Ubald to claim victory. Then I saw the rainbow, old and magnificent, arched in the sky above them and the church. In an instant I understood. The race could not end in the dark nave of a church, but would continue until a generation of racers passed beneath the shimmering arch of colors suspended overhead. The flickering saints were pagan gods aglow in the reddish light of dusk, and the rainbow was their aureola. The race was not to St. Ubald but to a timeless sanctuary, "and a rainbow was the open door of it."

It was an ancient rite and an ancient rainbow. They do not often occur any longer in our modern world. Once, long ago, they were more common.

—

BY A DEEP water hole in the desert land of Australia, an aboriginal bushman crouches with a smoking firebrand. Thirst has drawn him to the water's edge, but not without hazard. Lacking the powerful magic of fire and smoke, the bushman knows that an enormous many-colored serpent would rise from the pool's depths to drown and devour him. Afraid of the fire in his hand, the watery beast will leave the thirsty visitor unharmed. Throughout Australia the poolside creature bears the beguiling name "rainbow serpent" after its beautiful, multicolored body that occasionally raises itself over the earth to stand arched among the clouds. An ancient, powerful, and magical being, it has occupied the imagination of aborigines since their origins.

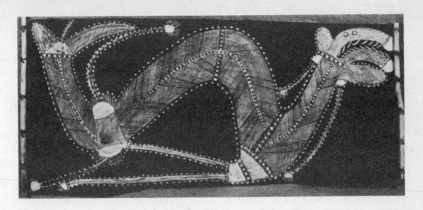

The Rainbow Serpent painted on bark by Namirrgi of the Dangbon Group on the Arnhem Land Plateau.

—

WHEN WANDERING NAVAJO gods came to a canyon chasm or river gorge, a rainbow bridge would overarch the abyss that they might pass easily through their lands. At their heels the sacred coyote would play, delighted by the newborn, rainbow trail. Some of the rainbow bridges solidified, becoming the natural rock bridges that today span ravines in the desert southwest.

—

"TO THE TROJANS there came as a messenger wind-footed Iris, in her speed, with the dark message from Zeus of the aegis."[2] Dark Zeus, master of the storm and thunderbolt, often called swift Iris to carry his messages to fellow immortals and to men. Iris, goddess of the rainbow, bridged the gulf of sky between the heights of Olympus and the battlefields before the walls of Troy. "Running on the rainy wind," as Homer speaks of her, I imagine Iris, in the form of a glorious rainbow, arching peacefully from stormy sky to a tumultuous plain carrying the thoughts of Zeus to Akhaian or Argive. The rainbow, in the archaic Greek mind, was a goddess and messenger—swift and changeable.

Rainbow—Daughter of Wonder

With the dawn of a sentient humanity, the spectral arc of the heavens we call the rainbow began its enchantment of the human mind. It has been a source of wonder, myth, and superstition, and has also been a phenomenon pregnant with implications for the developing scientific study of light.

Among the ancient Semitic peoples of the Near East, the rainbow marked the transition from a remote age of growing iniquity to our own, a theme that enters our own tradition through the book of Genesis. The rainbow's appearance to Noah sealed the covenant established between Jahve and "all flesh that is upon the earth." Never again would there be a destruction of the world by flood, never again would a sinful mankind be inundated and destroyed.[3] Seeing the rainbow in our half-lit sky can act as a comforting reminder of that pledge of God to man as we live precariously in a seriously threatened world.

To Homer the rainbow was a manifestation of the goddess Iris; from him we know her genealogy. Thaumas, the god of wonder, "chose Electra for his spouse / of the deep-flowing Oceanus child; / she bore swift Iris, fair-haired harpies too." In the Greek imagination Iris, the rainbow and messenger of the gods, united the deep-flowing outer sea called Oceanus that encircled the ancient world with the god of wonder, Thaumas. *Thaumas* was the Greek word meaning miracle. In the story of Iris's origins, the Greeks embodied their own sense of awe at the circling, colored rainbow; it was a miracle, a *thaumas*, one wed to water, to Oceanus. The rainbow is the offspring of water and wonder; Iris is the daughter of Oceanus and Thaumas. As

Plato wrote: "He was a good genealogist who made Iris the daughter of Thaumas."

In the early history of the rainbow, the arc of colors that stretched from heaven to earth became quite naturally a bridge that connected two worlds. As the Greek playwright Aristophanes wrote, "And Iris, says Homer, shoots straight through the skies with the ease of a terrified dove," bearing messages between gods and men. The motif was common in cultures other than the Greek as well. For North American Indians, Polynesians, and others, it was a pathway for souls to the upper world, in Japan the "floating bridge of heaven." In the Edda of Iceland, the god Odin in the form of Gangleri asked the way from heaven to earth and "Harr answered and laughed aloud: Now, that is not wisely asked; has it not been told thee that the gods made a bridge from earth to heaven, called Bifröst? Thou must have seen it; it may be that ye call it rainbow. It is of three colors, and very strong, and made with cunning and with more art than other works of craftsmanship." Closely guarded by Heemdel, the gods will make the last of their daily crossings over it at the time of their departure from our world, the "twilight of the gods." Then will the rainbow be destroyed.

The rainbow's mystery continues to enchant. Approach it and it recedes from you such that you can never walk beneath its vault, or reach its legendary end. What is the rainbow, woven of light and rain and eye? *Thaumas*, wonder, accompanies it throughout the centuries; Iris is as beautiful and miraculous now as she was to the ancient Greek. As a youth, Gerard Manley Hopkins captured the puzzle.

> *It was a hard thing to undo this knot.*
> *The rainbow shines, but only in the thought*
> *Of him that looks. Yet not in that alone,*
> *For who makes rainbows by invention?*
> *And many standing round a waterfall*

See one bow each, yet not the same to all,
But each a hand's breadth further than the next.
The sun on falling water writes the text
Which yet is in the eye or in the thought.
It was a hard thing to undo this knot.

The wonder felt by Hopkins or by Homer, which they shaped into poetry, is the root of philosophy, and the basis of science. Returning to Plato on Iris: "This sense of wonder is the mark of the philosopher. Philosophy indeed has no other origin, and he was a good genealogist who made Iris the daughter of Thaumas."[4] The genealogy of the Western mind reaches back to Iris, to man's wonder at the rainbow, and reaches forward to the twilight of the gods, and the destruction of the rainbow, the end of the covenant between Jahve and mankind.

The Phenomenon

It is not noon—the Sunbow's rays still arch
The torrent with the many hues of heaven....
Lord Byron

Where, when, and under what circumstances do rainbows appear? Inseparable from an understanding of a phenomenon is clarity about the details of its occurrence. We have all seen rainbows, but have you noticed in which direction they occur, at what time during the day, what the exact shape is, and how high they are in the sky? What are the colors of the rainbow and in what sequence do they occur? The answer to each of these questions is a hint about the cause of the rainbow.

While I was a research physicist at Boulder, Colorado, my wife and I were often pleasantly interrupted by rainbows at dinnertime. Small evening cloudbursts would quietly appear.

Coming off the mountains to the west, they would head out to the eastern plains that meet the Rocky Mountains at Boulder. The clouds and rain would pass overhead, obscuring for a moment the bright sunlight raying over the mountains at dusk. From our apartment we could not see the rainbow, but we knew it was there. We kept an umbrella by the door for rainbows. Going out back to the field behind the apartments, we looked out over the plains at the magnificent bows of colors hanging in the half-dark sky.

Perhaps my description has awakened a memory of a rainbow you have seen, and seen clearly. Now answer the questions with which I began. Where was my rainbow? Low on the eastern horizon. When was my rainbow? At evening. Already with these two observations we have the start of what I will call the "figure" of the rainbow, that is, its geometrical and temporal form. Once we get clear the figure of the rainbow, we can *see* the rainbow with understanding, as well as with our eyes.

Another feature of the rainbow's figure is the place of the sun. Where is it at the time of a rainbow? Virgil knew, and tells us.

> *As when the rainbow, opposite the sun*
> *A thousand intermingled colors throws*
> *With saffron wings then dewy Iris flies*
> *Through heaven's expanse, a thousand varied dyes*
> *Extracting from the sun, opposed in place.*

The sun stands opposite the rainbow. To find a rainbow, turn your back to the sun, and a thousand intermingled colors mark the path of dewy Iris's flight.

What are the number, kind, and order of the colors that comprise the figure of a rainbow? Virgil has told us a thousand of varied dyes. Aristotle asserted there are three: "the first and largest is red," with green and purple following as one moves inward. Xenophanes agreed: "This which they call Iris is also

a cloud, purple, and red and yellow-green [*chlorous*] to behold."[5]
The Icelandic prose Edda also held to three.

The invariant sequence of colors in the primary bow (from the inside out) are: violet, blue, green, yellow, orange, red. Their precise number is variable as they blend one into another, although often one sees three prominent hues.

What about the shape of a rainbow? Shelley responds:

> *From cape to cape, with a bridge-like shape,*
> *   Over a torrent sea,*
> *Sunbeam-proof, I hang like a roof,*
> *   The mountains its columns be.*
> *The triumphal arch through which I march,*
> *   With hurricane, fire, and snow,*
> *When the powers of the air are chained to my chair,*
> *   Is the million-coloured bow. . . .*

Aristotle correctly identified the shape of the rainbow as a segment of a circle that is never greater than a semicircle, at least under ordinary circumstances. Occasionally, when the position of sun, observer, and spray are just right, more of the rainbow may show itself. In his journal kept during the voyage of the *Beagle*, Charles Darwin recorded the sighting of a remarkable rainbow.

> The successive mountain ranges [of southern Chile] appeared like dim shadows, and the setting sun cast on the woodland a yellow gleam, much like that produced by the flame of spirits of wine. The water was white with the flying spray [of the *Beagle*], and the wind lulled and roared again through the rigging: it was an ominous, sublime scene. During a few minutes there was a bright rainbow, and it was curious to observe the effects of the spray, which, being carried along the surface of the water, changed the ordinary semi-circle into a circle—a band of prismatic colours being continued from both feet of the common arch across the bay,

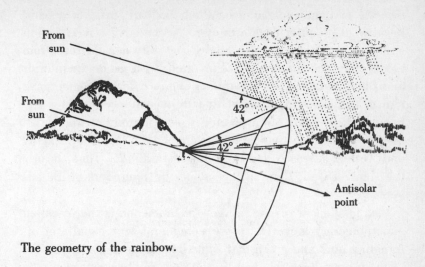

The geometry of the rainbow.

close to the vessel's side: thus forming a distorted, but very nearly entire ring.[6]

If its shape is usually semicircular, where is the center of a rainbow? It can be found simply as follows. Draw a line from the sun through the observing eye, continuing it to the ground under the rainbow. There you will notice a shadow, the shadow of your own head. The line connecting sun, eye, and shadow is the axis around which the rainbow circles.

A second line can be drawn from the eye to the rainbow itself. No matter where on the rainbow, and no matter whether in a spring shower or the mist of a garden hose, the angle formed between the first line and the second is always forty-two degrees—the rainbow angle.

A careful observer will also notice several other aspects of the rainbow. First, the rainbow appears like a boundary between a light-filled, interior space and a dark exterior band—Alexander's dark band. The colors of the rainbow thus arise at the meeting of light and dark. I called the dark exterior a band because at its outer edge a fainter, "secondary" rainbow often appears, again at a precise angle, now fifty-one degrees. The colors of the secondary bow are inverted, with red on the interior and blue outside. This perversion of the color sequence gave rise to the widespread belief that the secondary bow was the handiwork of Satan, who created it as a parody of the Lord's rainbow, which had marked his covenant with Noah. In Germany and Arabia, it was called the "devil's rainbow." Thus, beside the signs of God, there always stood the reminders of his adversary.

One final aspect of the rainbow phenomenon is the so-called supernumerary arcs. These are ephemeral arcs usually of alternating pink and green that sometimes appear just below the primary bow. Look for them in the next rainbow you see; they are often missed, but their graceful beauty rewards the attentive observer.

In Boulder, my wife and I would stand next to each other under the protection of our umbrella, each watching Iris run. Behind us the sun was setting, before us on the ground were the shadows of our two heads. A puzzle—reflect on it with Hopkins: "The rainbow shines, but only in the thought / Of him that looks." Two heads make two shadows, and so also two axes pass from sun through two pairs of eyes to the shadows of our heads. Thus do two rainbows circle the axes we each define. We "See one bow each, yet not the same to all, / But each a hand's breadth further than the next." Her bow and mine stood apart, as did we two. I would literally have had to see through her eyes to see her rainbow, and she had better close one eye so as to remove the final ambiguity. This was Hopkins's puzzle when writing on the rainbow.

All these features belong to the figure of the rainbow. In seeing the arc of colors before us, we should gather into the phenomenon the sun at our back, our eyes, the mist ahead, its colors and accessory bows. The geometry and timing of the rainbow together with these are all part of the godly figure of Iris.

Until the seventeenth century in France, the rainbow was known primarily as *iris* in honor of the Greek messenger goddess. René Descartes replaced that ancient name with a more prosaic and, until then, seldom-used phrase: *arc-en-ciel,* "arc-in-the-sky." This is one more instance where the transition from the mythic to the scientific reflects itself in the evolution of language. But evolution continues, and I wonder if the grandeur of the rainbow will not one day call for a name again filled with imagination.

Although displaced from her heavenly haunts, Iris still reigns in one part of our universe. Consider once more the line passing from the sun through the pupil of the observing eye. When that line is extended out to sunlit mists, a rainbow appears to encircle, like Oceanus, that axis of sight. But another far more modest ring of color encircles the same axis. It rests upon the aqueous organ of vision itself, between a black interior and a white exterior, at the threshold to an inner world, and "is the open door of it." That slight ring of color carries still the name of Iris—messenger of the gods.

Unweaving Rainbows

Let us now explain the nature and cause of halo, rainbow, mock suns, and rods, since the same account applies to them all.

Aristotle[7]

The history of the theory of the rainbow runs from Aristotle through Grosseteste, Descartes, Newton, and the wave theorists.[8] New concepts of optics either arose from careful study of the rainbow or were quickly applied to it in order to explain its mysterious features. The rainbow thus provides an ideal occasion for observing in a single natural phenomenon the changing eye mankind has brought to optical effects. It will reflect the evolution of light as we have followed it from remote antiquity to the end of the nineteenth century.

Our view of the rainbow began to change from the mythic to the scientific during the Hellenic period. Already we find an impressive treatment by Aristotle in his *Meteorology*, where it is considered a sublunar "meteor" of light together with halos, the aurora, comets, and meteorites. His observational account is remarkable for its clarity, and is the first to consider carefully the geometrical and physical basis for the rainbow. Aristotle submitted Iris to the same rigorous analysis by which Euclid had geometrized sight. As regards the physical origin of its colors, Aristotle treated clouds as made up of mist that acted as innumerable tiny mirrors, so small that they reflected the *sight* of the observer without maintaining the image of the sun (remember the visual ray of the Greek eye). Thus, from the eye a visual ray ran to the rainbow (i.e., to the mirrorlike droplets of a cloud), and from there to the sun. Aristotle's theory of colors viewed all colors as intermediates between white and black. Applying his theory to the rainbow, three colors arose: "When sight is relatively strong the change is to red; the next stage is green, and a further degree of weakness gives violet." These were the colors of Aristotle's rainbow.

The next great step in the understanding of the rainbow was given by Robert Grosseteste in his little book *On the Rainbow and the Mirror*.[9] He opens with thoughts regarding visual rays, and his metaphysics of light. He writes,

> We must not think that emanation of visual rays is just an imaginary idea without reality, as those persons profess who consider the part and not the whole. But we ought to know

that a visible *species* is a substance of like nature with the sun, which lights and radiates. The visible *species,* when conjoined with the radiation of an external illuminated body, completes perception. [10]

Grosseteste here summarizes beautifully the Platonic understanding of light common to the Middle Ages. The two species of emanation, from eye and sun, are like one another, and when they are conjoined, perception occurs.

From thoughtful considerations regarding reflection and the characteristics of rainbows, Grosseteste goes on to conclude that Aristotle's explanation of the rainbow as due to reflection of light from clouds is impossible. He suggests refraction instead. "It is therefore necessary that the rainbow be made by the refraction of the sun's rays in the moisture of a convex cloud." [11]

The phenomena associated with refraction are manifold and familiar. Whenever light (or for the Greeks, sight) enters or exits a new medium at an angle, it appears deflected at the interface. Grosseteste thought that perhaps the moisture of a cloud, when specially shaped into a convex form, might refract the light of the sun into a rainbow. All details of how this would occur are missing, but Grosseteste had advanced a seminal idea that would prove correct.

Approximately a century after Grosseteste's little treatise, Theodoric of Freiberg (not Boethius' murderous lord) took matters a step further. Sometime after 1304, Theodoric performed a study of the rainbow that was revolutionary in method as well as results. His predecessors had considered the watery medium of the cloud as a whole, neglecting the role of each tiny raindrop. By contrast Theodoric considered the cloud to be made of individual drops each of whose interaction with light was important. If a single drop could be studied with sufficient precision, then by compounding the effects of many similar drops, one could build up a rainbow. This is just the principle of analysis that was later used so effectively by Newton in the calculus and the problem of gravitation.

The small size of a water drop made its study difficult, so Theodoric enlarged it to a spherical glass vial of water placed in the sun. From his investigations, he concluded that as light enters the raindrop it first refracts, is then reflected at the concave inner surface at the rear of the drop, and then exits, once again with a refraction. "Such radiation, I say, serves to explain the production of the rainbow," wrote Theodoric.[12]

As if this were not enough, Theodoric went on to consider the production of the fainter secondary bow. He rightly recognized that an exactly similar process takes place but now with an additional reflection inside the droplets.

Finally, in putting the droplets back into the sky, Theodoric realized that each drop can offer only a single color to the eye depending on the precise geometrical relationship between eye and drop and sun. Thus a rainbow, seen all at once, must be the product of many such drops. "Consequently, if all of the colors are seen at the same time, as happens in the rainbow, this must necessarily result from different drops which have different positions with respect to the eye and the eye to them."[13]

Look at dew in the morning sun. Each droplet sparkles like a gemstone. Find one that shines brightly, and slowly move your head up and down. The glistening colors you see always pass through the same ordered sequence of the rainbow: red, yellow, green, violet. Place these dewdrops in the sky as rain, and they continue to glisten, each with a single color. Move your head again and new droplets shine red while the old ones turn yellow. Geometry is everything in the figure of light. Colorless in itself, each droplet sparkles, a diamond of colored light standing between the eye of Horus and the eye of man.

—

OPTICAL THEORY GRADUALLY encircled the rainbow, but two final contributions were still needed. The first was provided by René Descartes and the second by Isaac Newton.

Descartes's contribution to the rainbow was subtle but profound, uncovering the reason for the rainbow angle and also the presence of Alexander's dark band. By way of introduction, consider for a moment the trajectory of water from your garden hose. Slowly vary the angle of the nozzle from straight forward to directly up, and note the changing distance the water travels. Initially, as the angle increases, so does the distance. However, when the nozzle is at a forty-five-degree angle, the water reaches its farthest; increasing the angle still more only causes the water to retreat. As Galileo first noted for artillery, maximum range is achieved at forty-five degrees, with shorter distances being traveled for angles both greater and less than that.

In studying the rainbow, Descartes deduced a similar behavior. As a codiscoverer with Snell of the first proper mathematical law of refraction, Descartes had all the theoretical tools (but one) for a modern account of the rainbow.

We can follow Theodoric's good example, as Descartes also did, and use a large spherical vial of water to simulate a raindrop. Send into it a single narrow beam of white light. Allow the beam first to enter the sphere at its center.

Slowly and steadily slide the entering beam across the water sphere toward its edge. The beam emerging from the vial correspondingly moves farther and farther to one side until (as with the hose) it slows to a halt and doubles back. At the moment it halts, the refracted and reflected beam bursts into color—the colors of the rainbow. The angle between the incoming and outgoing beams is then just forty-two degrees—the rainbow angle. "Doubling back" is the key to the rainbow. Without it, no

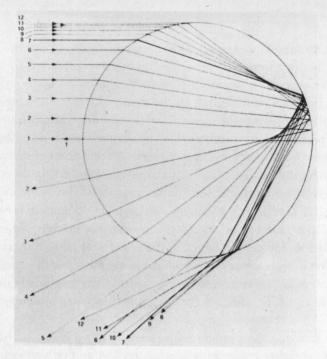

Light rays entering a raindrop (above) and leaving (below). Notice how they crowd around ray seven. Here is where the rainbow will appear.

amount of refraction would be sufficient to produce a rainbow in the sky.

As an additional bonus, the dark band receives an explanation. Just as no water falls beyond the maximum range of the hose no matter at what angle you hold the nozzle, so also no light is refracted beyond the rainbow angle no matter how light hits the droplet of water. This absence of light implies a dark region beyond forty-two degrees—Alexander's dark band.

The secondary bow just outside the dark band finds a similar explanation. The additional interior reflection present in the latter has the consequence that the emergent beam approaches a new fifty-one-degree rainbow angle but now from the *opposite* side. The refracted and reflected rays halt as before and retreat back to larger angles yielding colors in inverted order. Alexander's dark band is preserved in the space between forty-two and fifty-one degrees, where neither can reach, a dark abyss between the bow of Jahve and that of Satan.

Descartes had presented the complete geometry of the rainbow, but it was entirely monochrome. He lacked a reasonable explanation for the essence of the bow: its colors.

—

OVER TWO THOUSAND years after Homer, in 1672, Newton, then still a solitary genius at Cambridge University, sent a letter to the Royal Society of London. The content of that letter was an outline of a new theory of light and color that, within a century, would complete not only our scientific understanding of the rainbow, but change our imagination of it as well. The rainbow, he suggested, was composed of "rays diversely refrangible," each one of which induced the sensation of a specific color.

Recall Newton's corpuscular theory of light. White light, according to him, is not at all what it appears to be. In fact it is made up of rays or atoms of light of several kinds, each kind capable of causing a different color sensation. Newton made the significant experimental observation that lights of different colors are refracted by slightly different amounts. The import of this for the rainbow is great. Although Descartes could account for the rainbow angle as the extreme angle prior to return, there remained still the question of the colors of the rainbow. Descartes understood the geometry of the rainbow in black and white only. Newton's theory granted it color. The rainbow angle is slightly different for every one of the rays that together compose the white light of the sun. That is, the extreme angle prior to return changes systematically from red through green to violet. This is called dispersion. Regardless of the correctness or falsity of Newton's corpuscular theory, the fact of dispersion was the last missing piece. The mechanism of dispersion, its physical cause, is not what Newton imagined it to be. However, his account, together with its implied corpuscular dynamics, carried the day nonetheless.

The eighteenth century rejoiced. At long last the rainbow had given up her secret, the goddess Iris was finally despoiled by Sir Isaac Newton, a bizarre irony for a man who had so little use for women. The poets sang Newton's praises, among them James Thomson:

> *Even Light itself, which every thing displays,*
> *Shone undiscovered, till his brighter mind*
> *Untwisted all the shining robe of day;*
> *And, from the whitening undistinguished blaze,*
> *Collecting every ray into his kind,*
> *To the charmed eye educed the gorgeous train*
> *Of parent colours.*[14]

Light itself was now illuminated by the even brighter light of Newton's mind, and stood revealed as a train of parent colors normally twisted into whiteness. In prism and rainbow the "yellow tresses" of sunlight are disentangled and displayed as rays diversely refrangible.

With the rainbow now explained, only a few minor curiosities remained, such as the supernumerary bows of pink and green just below the primary bow. The modern account understands these to be due to interference effects of the kind first suggested by Thomas Young. Thus do we find in the rainbow a chronicle of every feature of optical theory, old and new.

Around it also has swirled the emotions of Romantic hearts who saw in the triumph of optics over the rainbow the death of poetic sensibilities. Mark Twain wrote of his beloved Mississippi that he could see her two ways: as writer or as riverboat pilot. Through the eyes of the writer, every swirl, fallen bough, or lapping wave was the occasion for prose redolent with allusions and feelings. To the calculating eyes of the pilot, these very same signs were the indicators of shifting sandbars, the telltale effects of a recent storm, or some other natural occurrence that could threaten the safe passage of his vessel. Are these two views not completely incompatible? While one reigns the other sleeps, like day replacing night and night day.

Those who lived under the inspiration of night sent up a lament for the loss of the rainbow.

Awful Rainbows

> *...Do not all charms fly*
> *At the mere touch of cold philosophy?*
> *There was an awful rainbow once in heaven:*
> *We know her woof, her texture; she is given*
> *In the dull catalogue of common things.*
> *Philosophy will clip an angel's wings,*
> *Conquer all mysteries by rule and line,*
> *Empty the haunted air, the gnomed mine—*
> *Unweave a rainbow....*
>
> John Keats, *Lamia* (1820)

Some, with Alexander Pope, would eulogize Newton, declaring, "Nature and nature's laws lay hid in night, / God said, 'Let Newton be!' and all was light." To them, the development of scientific consciousness was progress, and its accomplishments distinguished our time from that of earlier barbarism and ignorance.

Others, with Keats, lamented the loss of Iris—swift and beautiful—from the heavens. They saw the advent of science as hastening the twilight of the gods, the destruction of the Bifröst bridge. The rich, inner soul-world of the rainbow was plundered, and in its place streamed abstract atoms of color. No matter how much finery the poet might hang on Newton's text, its lifeless bones shone through. A past golden age seemed lost forever. In response they often turned their backs to the new scientific imagination of nature.

The painter Benjamin Robert Haydon included in his diary for December 28, 1817, a description of an "immortal dinner party"[15] at which Keats, Wordsworth, and Lamb were present. Keats, in a highly humorous way, abused Haydon for including Newton's head in his painting: "a fellow who believed nothing unless it was as clear as the three sides of a triangle." In his account of the party, Haydon told of how Lamb and Keats "agreed that he [Newton] had destroyed all the poetry of the rainbow by reducing it to its prismatic colours. It was impossible to resist Keats," and they all drank "Newton's health, and confusion to mathematics."[16] Their attitude was captured by Thomas Campbell in his poem on the rainbow.[17]

> *Triumphal arch, that fill'st the sky*
> *When storms prepare to part,*
> *I ask not proud Philosophy*
> *To teach me what thou art.*
>
> *Can all that optics teach, unfold*
> *Thy form to please me so,*
> *As when I dream of gems and gold*
> *Hid in thy radiant bow?*
>
> *When Science from Creation's face*
> *Enchantment's veil withdraws,*
> *What lovely visions yield their place*
> *To cold material laws!*

From a consciousness that saw in the rainbow a covenant with God—or a goddess herself—we have come to see the rainbow as an ephemeral phenomenon of mundane light manufactured by refraction and rain. We have exchanged an ancient view or imagination of the world for the one of contemporary science, and in doing so we have, in Keats's words, "Unwoven the rainbow."

The shift in view is a simple fact. This may be seen as progress by some and a fall from grace by others, but we simply do inhabit a different world from that of the Australian aborigine or ancient Sumerian. It differs less because of technical advances in our external world than because of a revolution in our way of thinking and seeing, a revolution located within us. We are the inheritors not only of an external culture but of an internal one as well. In fact these two are inextricably entwined, one with the other. External history is inseparable from the history of our interior landscape, and so the history of the rainbow will be a history of the mind seen as if through one facet of a many-sided gem.

The history of the rainbow from the age of myth to contemporary optics is an exemplar wrought in miniature of our unfolding penetration of and relation to natural phenomena. It is a powerful story that moves from Iris, messenger goddess of Homer's Greece, through the philosophers of antiquity, to Descartes, Newton, and Young, who laid the foundation for a modern understanding of the rainbow.

Yet while engaging in itself, the history of the rainbow hides within it another story far more significant than an external history of science. For the changing images of the rainbow reflect to us momentous changes in the fabric of consciousness itself. The history of light, the rainbow, and more generally the history of science continue to act as a text in which we read the psychogenesis of the mind.

Using Your Own Light

Seeing into darkness is clarity.
Knowing how to yield is strength.

Use your own light
and return to the source of light.
This is called practicing eternity.
Lao-tzu[18]

Owen Barfield begins his brilliant little book *Saving the Appearances* by reflecting on the rainbow.[19] His question is simply: *Is it really there?*

Since you cannot approach a rainbow, cannot touch it, smell it, get under or inside it, is it really there, or is it only a special scattering of light from tiny fluid globes? One consideration in favor of its reality: We have all seen rainbows. As a civilization we share the experience, and so whatever rainbows really are, they exist as a collective phenomenon.

But then, Barfield asks, consider a tree. Accord to it the same treatment you did the rainbow. It certainly differs from a rainbow in that you can approach it, touch it, perhaps even smell it. You still cannot get inside or easily under it, but its reality seems hardly an issue. Yet ever since Galileo, and especially in our own century, physics has insisted that the tree, like the rainbow, is made up of small globes of matter called atoms that scatter light according to particular laws. Some small fraction of that light makes its way to your eye and you "represent" the tree to yourself by an unknown and miraculous process.

Quivering aspen leaves show us their silvery undersides as dark green pines gently rock in the same breeze, creaking their higher branches. Perhaps an evening shower has passed overhead and a rainbow hovers above the treetops and before a distant hill. Physics may say that this is all a complex electromagnetic interaction among neighboring cellulose molecules, water, their constituent atoms, and light, but we see neither light waves nor atoms; we see, hear, smell, and feel trees and rainbows. They and we are the familiar objects of consciousness.

We care for such things, and to such common things we are willing to give ourselves.

Children, trees, and rainbows exist. About this neither Barfield nor I wish to argue. What they are in themselves, in their essence, we may not fully know, but we do "represent" them to ourselves in robust colors and character. And strangely, our entire contemporary culture represents them roughly as we do. If we take to seeing rainbows and trees when others do not, our hallucinations may land us in a psychiatrist's office. Thus the peoples of every time and place shared representations. Barfield calls them "collective representations."

Now comes the critical two-part question: How do such representations arise, and can they differ from one place or time to another? Think back to our discussion of vision. Recall the puzzle of the interior ray, the light that spreads from the lamp of the body, our eyes. From the case of S.B. and others born blind, we learned that seeing entails far more than the possession of an operational sense organ. Into raw sensations flow things such as memory, imagination, mental habits, feelings, and even our will (in as much as we *attend* to something). Without the light that we bring to sensations, the world is meaningless and dark. If the physical eye is a camera obscura, a "dark chamber," then sight requires, in addition, a spiritual eye for which Empedocles' image of the lantern is a more fitting metaphor.

Modern evolutionary biology quite naturally sees the present-day physical eye as the end product of a long evolutionary development. Eyeless fish swim the dark, lightless waters of Echo River in the Mammoth Caves of Kentucky. Lacking light, the fish evolved no eyes. By contrast, light has bathed our planet for millions of years. Under its persistent influence, physical organs of perception have arisen suited to it. Is it not also possible that over the tens of thousands of years that intelligent humankind has existed, a similar evolution might have taken place in the development of the spiritual capacity that grants meaning to the proddings of our senses? Perhaps as Goethe says: "The eye owes its existence to the light. Out of indifferent animal organs the light produces an organ to correspond to itself; and so the eye is formed by the light, for the light so that the inner light may meet the outer. . . . If the eye were not sunlike, how could we perceive the light?"[20]

If, over millennia, the creative outer light of the sun has called forth our physical eyes, what were the forces that crafted, kindled, and colored our inner light? As the English chemist and philosopher Michael Polanyi puts it, our language, tools, and action create faculties: ". . . we *interiorize* these things and *make ourselves dwell in them*."[21] By dwelling in them, new organs of cognition arise. Musing on the history of light and the rainbow, I think we can begin to sense those changeable psychic forces in which man has lived, and that have granted meaning to experience.

In our study of light, we have been concerned not with the "right" explanations of contemporary optics, but with our changing imaginations of light. Accord to each its rightful power and glory, for they reigned long and well in their time. Nor did the electromagnetic view of Maxwell have a longer tenure; we have not yet gotten to the quantum theory of light.

It would be better if we asked not, were prior views of light *true*, but rather, what is the significance of the view? What lived within the soul of Egyptian or Greek that was content with the eye of Horus and swift-winged Iris? As John Stuart Mill wrote, paraphrasing Coleridge, "The long duration of a belief is at least proof of an adaption in it to some portion or other of the human mind; and if, on digging down to the root, we do not find, as is generally the case, some truth, we shall find some natural want or requirement of human nature which the doctrine in question is fitted to satisfy."[22]

Our conscious awareness of light, the rainbow, and, indeed, all phenomena has changed vividly during the thousands of years that we as a species have viewed them. As humans have evolved biologically, so, too, has human consciousness evolved. Moreover, that evolution continues in our own day, and its future remains open. What will we make of it? I wonder.

Some, rejecting the evolution of mind entirely, will hold fast to a nineteenth-century materialistic imagination of light in an age whose science cannot support them. They may speak of "photons" but are actually only revisiting Newton's corpuscular theory of light. Others, better informed, will adopt a radical relativism. All knowledge is purely conventional, they will say. To ask after the nature of light is but a socially sanctioned game; the concept of light itself is merely a convention. Their deconstructive enterprise is useful in that it undermines the complacency of science, but in the end it offers nothing in its place. The end result of deconstruction is a pile of ruins: nihilism. Evolutionary change is accepted by practitioners of this view, but it (evolution) is ultimately meaningless.

The third avenue is one seldom taken because it demands the most. It requires that we imagine light, ourselves, and the world as richer and deeper than the mechanical imagination can capture. Goethe, among others, advocated such an understanding. With care, but not fear, they sought a way that the physical might open out on the spiritual. The future consciousness necessary to accommodate their science could not be mechanical merely, but always ready to dwell lovingly in new phenomena, actions, and expressions in order that fresh faculties of cognition be created. The essence of light would not be a physical thing, an idol, but a spiritual reality or agency.

As we enter the twentieth century in the pages ahead, all three avenues are taken. The first two, materialistic realism and various forms of relativism, are easily treated because many scientists and philosophers have worked these fields already. By contrast, the modern spiritual history of light is less familiar, and seems, at first, antithetical to its scientific treatment. Yet

I think this view mistaken. The torch carried by Goethe and others such as Coleridge and Emerson, was picked up by the philosopher and "spiritual scientist" Rudolf Steiner. Therefore, as we watch the quantum theory of light unfold through the struggles of Planck, Einstein, and Bohr, we will also follow the simultaneous creation of a modern spiritual imagination of light by Rudolf Steiner.

—

THE BIOGRAPHY OF light is like the rainbow; a complex harmony of many features that weaves together the rigorous figures of natural law and the changing human soul to create a transient appearance of colors. The sun low at one's back can write into the mist before us the text of two majestic bows—one the Lord's and the second Satan's—separated by a chasm of darkness. Graceful pink arcs work their way into the light-filled space beneath the brighter primary bow. The unpretentious intricacy of the rainbow easily becomes a metaphor for human life. The mythic and scientific mingle still within us.

Goethe's Faust, when broken and dispirited, was gently awakened by the sounds of Aeolian harps and the song of the spirit Ariel. In the watery vapors of a nearby cataract, he saw hovering a glorious rainbow. His words capture all the tumult of the waterfalls and his anguished soul, his dark thoughts, and the tranquil radiance of sun and human spirit. Between these two flowers the rainbow whose many-hued reflections mirror life's aims and actions.

> *Nay, then, the sun shall bide behind my shoulders!*
> *The cataract, that through the gorge doth thunder*
> *I'll watch with growing rapture,'mid the boulders*
> *From plunge to plunge down-rolling, rent asunder*
> *In thousand thousand streams, aloft shower*
> *Foam upon hissing foam, the depths from under.*
> *Yet blossoms from this storm a radiant flower;*

Door of the Rainbow

The painted rainbow bends its changeful being,
Now lost in air, now limned with clearest power,
Shedding this fragrant coolness round us fleeing.
Its rays an image of man's efforts render;
Think, and more clearly wilt thou grasp it, seeing
Life in the many-hued, reflected splendour.[23]

CHAPTER 8

Seeing Light—
Ensouling Science:
Goethe and Steiner

The purest and most thoughtful minds are
those which love color the most.

John Ruskin

Imagine that your laboratory was on the grounds, and amid the struggles and joys, of a hospice run by Mother Teresa. In that context, what would be the important questions concerning light and color? My introduction to color science occurred in such a setting, under the tutelage of an unusual and wonderful man— Michael Wilson.

The year was 1974; Wilson and I were driving from Exeter University to his color lab in Clent, near Birmingham, for a week of work together. As we wound along the Atlantic coast of south England, he pointed out the misty beauties of heather and sea, but the mountains of Wales, which he hiked at every opportunity, were his real love. There, sky and stone, pasture, sheep and shepherd, mingled on blue-green hills. Forty years my senior in science, Wilson was a master of color; I was a young American graduate student more accustomed to electron-impact excitation of helium than the rainbow's hues. I had come on the advice of a trusted professor, knowing Wilson had been chair of the Color Group of England and author of several important scientific papers on color. I knew less about his private life, but in the days that followed, and during the years of friendship that ensued, I would learn just how closely his life was connected to light and color.

From the moment we drove onto Wilson's estate, I knew my introduction to color was to be strikingly different from university physics instruction in Ann Arbor, Michigan. Leaving the VW van by an old farmhouse, we walked a path up to Sunfield Children's Home. Some thirty years before, Wilson and a few dedicated friends had founded a home for mentally handicapped children that had since grown into a large residential community. Climbing the hill that overlooked Sunfield's property, we looked out over fields, trees, and groups of obviously disabled children with their attentive staff. I suddenly realized that I would study color here, within this community of handicapped children, and that this same community had surrounded and supported the color work of Michael Wilson. I had become accustomed to "centers of excellence," staffed entirely with brilliant, high-strung scientists and technicians. At Sunfield I found another tempo and style of investigation, one guided by compassion rather than grant deadlines or scientific rivalry. The setting of Sunfield encouraged me to adopt a different attitude to my scientific work, one I have struggled to maintain ever since.

On our way back we stopped to visit with staff and children in the main building. It was then I began to discover the full range of Wilson's engagement with color, for not only had he worked scientifically with the straight formalism of modern color theory, but for years he had also sensitively experimented with color as a therapy for the handicapped children of Sunfield.

Clearly, here was a man of independent mind and warm heart, one who commanded the respect not only of scientists on both sides of the Atlantic, but who also received, and returned in kind, the love of the Down's syndrome, spastic, and autistic children of Sunfield. Wilson's love of color had not only purified his mind, as Ruskin had predicted, but it had also bridged worlds I had always seen kept severely apart. I longed to know, what were light and color to this man?

From Land to Goethe

In November of 1957, Edwin Land (the inventor of instant photography) lectured on color, with demonstrations, to the National Academy of Sciences and to the Rockefeller Institute for Medical Research.[1] His presentations—widely reported in the press—had startled the scientific community. In them Land challenged the very foundations of contemporary color theory. Six months later, Land lectured to the Royal Photographic Society in London, and shortly thereafter Wilson began his well-known study of Edwin Land's radical revision of color science.[2]

During my visit, Wilson led me through Land's demonstrations, one by one, in his well-equipped color laboratory (located in a converted set of farm sheds). Land's demonstrations were truly astonishing. Nothing I had learned at the university could explain what I was seeing. The standard basis for understanding color had been laid down first by Newton. With the subsequent advent of the wave theory of light, the connection between color and wavelength then became commonplace. Together, these formed the orthodox framework for the understanding of color. Although sufficient for the rainbow, with them alone one simply could not make sense of what I was seeing. Land's experiments seemed to challenge the scientific notions of color with a greater force than any previous experiments.

—

NEWTON HAD SHOWN that if one extracted, say, yellow light from the spectrum produced by a prism, and mixed it with orange light similarly produced, then a color intermediate between the two—a yellow orange—appeared. Its particular hue depended on which color dominated the mixture, orange or yellow. Land performed the same experiment but with a single important modification. He projected the yellow and orange light beams through black-and-white photographic transparencies. The transparencies depicted an identical still-life scene but photographed through different-colored filters. With only the yellow image projected, one saw a purely monochrome-yellow still life on the screen. None of the original colors of the scene were present, only shades of yellow. The same was true when the second image alone was projected through the orange filter. Now, however, the still life was entirely in shades of orange.

With Newton in mind, what would you expect to see if both images were projected on top of one another? Hues somewhere between yellow and orange as before? That is what I expected, and most of the members of the National Academy of Sciences expected the same. However, you do not see yellow oranges, far from it! Reenacting Land's demonstrations with Wilson, I saw what appeared to be a full range of colors, including reds, blues, and greens. But these were colors I "knew" simply could not be there! My eyes told me one story, my training as a physicist told me another. What was going on?

Michael Wilson answered my questions in his book-crammed study, our feet warming near one another at an electric hearth. In doing so, he took me back to the dawn of the nineteenth century, and to the German poet Goethe. Ironically, the basis for understanding what had shocked the National Academy of Sciences was to be found in the color studies of Germany's great literary genius. Now I understood the name of Wilson's institute: the Goethean Science Foundation.

We can join Goethe at the moment of his entry into the story of light. In doing so, we witness the entry of a new vein into light's history, one that carries us away from the mechanical and electrical imaginations of light and toward a renewed spiritual conception of it.

Colors of the Eye

We reserve the right to marvel at color's occurrences and meanings, to admire and, if possible, uncover color's secrets.[3]

Goethe

In January of 1790, Germany's great literary genius, Johann Wolfgang von Goethe, lifted a prism to his eyes hoping to glimpse the secret of color. The prism had been borrowed long before, along with other optical equipment, from privy councillor Hofrat Büttner of Jena, but now a messenger stood impatiently at the door insisting upon its immediate return. The owner had understandably despaired of ever getting back the equipment, all of which had languished untouched for months in Goethe's closet. Goethe realized that Büttner's importuning could be put off no longer. The box of optical equipment was retrieved, but before handing it over to the messenger, Goethe could not resist pulling out a prism. He wished to see, if ever so briefly, Newton's "celebrated phenomenon of colors" known to him from childhood.

The prism before him, and Newton's theory of light firmly in mind, Goethe looked at the white walls of the room expecting them to be dressed in the colors of the rainbow. Instead he saw only white! Amazed, he turned to the window, whose dark cross-frame stood out sharply against the light gray sky behind. Here, at the edge of frame and sky, where light and darkness met, colors sprang into view in the most lively way. Like a shot sounding in the mountains, Goethe realized and spoke aloud, "Newton is wrong!" Returning the prisms was now unthinkable, so Büttner's servant was, yet again, sent away empty-handed. Goethe's study of color had finally begun.[4]

For the next forty years, Goethe experimented with light and color, seeking not only color's secrets, but also a method of investigation more congenial to his temperament, one that was at once objective and yet devoted to nature, simultaneously science and art.

At the end of his long life, Goethe looked back and declared at least modest success. In his own estimation, his scientific studies ranked above all else, greater than *Faust*, his poetry, plays, and novels. All these, he said, would dim in the luster of other poets. Less perishable would be his contributions to science, and foremost among them would be his massive study of light, darkness, and color. To his secretary Eckermann, Goethe often remarked in his last years that "I do not pride myself at all on the things I have done as a poet. There have been excellent poets during my lifetime; still more excellent ones lived before me, and after me there will be others. But I am proud that I am the only one in my century who knows the truth about the difficult science of color."[5] This was Goethe's ripe judgment spoken after fifty years of effort in botany, color, zoology, geology, meteorology, and many other areas of scientific endeavor. His words have troubled a great many of his admirers. His singular stature in the history of world literature is unquestioned, but his science has remained an enigma. Was the great man duped, or have his efforts been consistently misunderstood?

—

IN ORDER TO probe the character of Goethe's genius, we must learn to see the world partly through his eyes. For his proposal is not given in the idiom of conventional science and so is easily misread. Goethe does not advance a competing theory of light in the usual way, but rather offers a wholesale reinterpretation of the scientific enterprise itself. He does so not abstractly as a philosopher, but concretely as an artist-scientist, or as Emerson would later speak of him, as the paradigmatic "poet-savant." In Goethe, imagination joins with experiment in a way that brings us suddenly much closer to light.

Büttner's impatient messenger had nudged Goethe into a muddled reenactment of Newton's prime experiment, and so into color science. Yet the peculiar manner of his own entry into color science was, Goethe felt, inappropriate for others. Characteristically, he chose to introduce his readers to color with precisely those phenomena long dismissed as inconsequential or outright misleading—optical illusions.[6]

—

ONE EVENING WHILE at an inn, Goethe noted the entry of an especially beautiful young woman. Her white face was radiant, contrasting strongly with her jet-black hair. A scarlet bodice shaped her ample figure to advantage; Goethe was fascinated by her. From across the room, he watched intently as she stood in the light. After some moments, she stepped away, but as she did so a striking twin appeared on the white wall behind, exactly where she had been. Now, however, a black face appeared, surrounded by a bright nimbus of light. Instead of a scarlet bodice, the dark beauty was dressed in magnificent sea green. The specter was even more lovely than the visitor.

We have already met and puzzled on similar perceptual conundrums, but in Goethe's three-part treatment of light and color, *Towards a Theory of Color*, they take center stage. Instead of dismissing them as mischievous phantoms of the mind, Goethe informs us at the outset that these very phenomena will be the basis for his entire theory. Indeed, for him there is no such thing as an optical illusion: "Optical illusion is optical truth!"

he defiantly declares.[7] In them is evidenced the living inter-action of our inner nature with outer nature. Of especial value are pathological instances of color experience, for through them the truths of color and cognition are most clearly rendered. No wonder Goethe anticipated Dalton in the study of color blind-ness.

It was a brilliant stroke. At the outset Goethe distinguished himself by advocating a radically different approach to color and light, which, true to his poetic genius, was literally founded on the imagination. What had been mere illusions became sign-posts on the path to truth. Phantoms became facts, and through them the subtleties of color vision could slowly be unraveled.

In Goethe's wake followed a stream of notable scientists. They self-consciously followed his lead by studying illusions and path-ologies of perception. Edwin Land's presentation to the National Academy of Sciences, as well as the fascinating studies of Oliver Sacks, are in the same tradition and ultimately rest on the opening paragraph of Goethe's *Theory of Color*.

—

AMONG GOETHE'S PAPERS is preserved a simple portrait of a woman, but in reversed colors; perhaps the same striking figure he saw that evening in the inn. Staring at it for a minute, and then slipping it to the side, one sees the original fair-skinned visitor reappear, hovering phantomlike over the ground behind it. From experiences such as these, Goethe fashioned his ap-proach to light, one that starts with the eye's own world of colors before going on to investigate those of outer nature.

Rekindling the Empedoclean Fire

How do we see physically? No differently than we do in our consciousness—by means of the productive power of

> *imagination. Consciousness is the eye and ear, the sense*
> *for inner and outer meaning.*
>
> Novalis

In mundane imitation of the tavern scene, place a small brightly colored object before you on a white sheet of paper. Stare fixedly at it for thirty seconds or so. Now slide the object away, or avert your eyes to a neutral-colored surface, and relax. Before you hovers an image whose shape is identical with that of the viewed object, but whose color is different from, in fact opposite to, that of the original. It may wander and fade, but can be renewed for a time by blinking. Its edges change, passing through a color sequence, growing inward, and then fading from sight altogether. If you have done the exercise, then you have seen a *negative afterimage*.

Since Aristotle first described them in *On Dreams*, afterimages have drifted in and out of the literature concerning sight. Bright objects induce dark afterimages, dark objects induce light ones; a red object provokes a green image, while a green one calls forward a red afterimage. In these phenomena we sense a lawful pattern, a truth about sight. Goethe called it the "law of required change."[8] The eye "is compelled to a form of opposition: setting extreme against extreme ... quickly merging opposites and striving to achieve a whole." The pattern of appearance can be elegantly represented by the color circle.

When stimulated by a color taken from one part of the color circle, the eye seeks "to complete the circle of colors within itself" by providing its opposite.[9] This law takes on all the more significance when the opposite color appears not only subsequent to the initial impression, but *simultaneous* with it, as is the case in the phenomenon of colored shadows, also carefully studied by Goethe. While somewhat difficult to follow in words, the effect is striking, and I encourage you to try it for yourself.[10]

First noticed by Otto von Guericke (the inventor of the vacuum

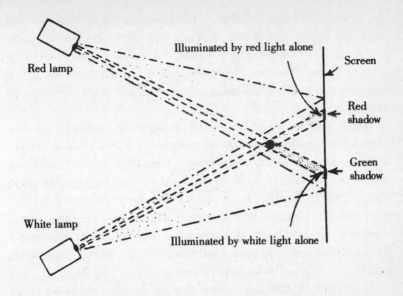

Red lamp

Illuminated by red light alone

Screen

Red shadow

Green shadow

White lamp

Illuminated by white light alone

Colored shadows appear when a region is lit by two differently colored sources. How can the green shadow arise without green light?

pump) in 1672, colored shadows appear whenever two differently colored sources of light illuminate a single region, as shown in the diagram above. Clearly, two shadows can form, one caused by blocking the light of each source. Allow one source to be uncolored, for example, and the second to be red. Then a little thought shows that one shadow is illumined only by red light and the other only by white light. There are, therefore, three regions on the screen: red alone, white alone, and red plus white. What do you expect to see? Red alone gives a bright red region, which makes sense. Red and white together give a pale red, which also meets our expectations. But much to everyone's surprise, white alone is seen as green! Neither source of light itself is green; no green anywhere, and yet one sees an unambiguous green.

Here once again is the "law of required change." When the

eye (understood now as the entire visual system) is under the dominant impression of red, as it is in the above, then it responds by "seeing" white as green, which is the color complementary to red. If one changes the red source to blue, then the white region is seen as yellow orange, the complementary of blue, and so on.

These phenomena of "chromatic adaptation," as they are now called, demonstrate to us the important truth that sight is active, and possesses its own laws. Edwin Land's color investigations can only be understood if we include this active dimension to sight. Prior to Goethe, such effects had been dismissed as distractions. For Goethe, by contrast, such illusions were the proper starting point and basis of a scientific study of color. Especially through illusions, we gain a glimpse of the character of sight, and of the "ideal," imaginative power active within it.

A student of Goethe, as well as of modern color science, Wilson was especially qualified to analyze Land's experiments. Wilson demonstrated that the blues seen in the Land experiments were caused by the same process responsible for colored shadows, so carefully studied since Goethe's time. The human visual organization accommodates to the dominant color, yellow orange in Land's case. The lack of this color—like a colored shadow—will be seen as blue violet, the complementary color to yellow orange. The context, as well as the object itself, strongly influences the color we see. All the inexplicable colors in Land's presentations derive from our unconscious adaptation to the color context.

In the work of Goethe and Land, the mobility and intelligence active in sight is drawn sharply to the fore. One cannot neglect the power of the eye, its interior light, as Empedocles would have said, if we would understand our experience of color. The color "illusions" of Land and Goethe provide us with a rare glimpse into the ubiquitous presence of mind in vision. Our every perception is literally colored by context, prior experience, indeed, by every aspect of our inner world. These are all active in producing color. As Oliver Sacks remarked after his study of the color-blind painter, color is an integral part of ourselves, of our lifeworld. As such, we affect it, but it also affects us, far more deeply than we realize.

—

THE CHILDREN HAD just left. Their watercolor paintings, rich with colors, were still glistening in the afternoon light. The most able children of Sunfield had finished a session with their art teacher. She had worked according to a therapeutic plan developed in collaboration with the home's doctor, the class teacher, and Michael Wilson. A pale yellow crescent moon hung in the center of each painting surrounded by a dark blue sky.[11] The children were drawn into the radiant experience of light, and the embracing mood of the night sky.

At Sunfield, all those who might benefit from it were given color therapy. Children unable to paint were treated in a color projection room or color pool. My own session with Ursula Gral, the color-room therapist, convinced me of the subtle but powerful effects of color.

Sitting me in a small room, Ursula gradually and silently led me on a journey into color space. The room lights changed color, dimming to a deep blue. When they were at their dimmest, a curtain opened before me and colors (projected from behind the wall) delicately filled the large screen, the room, and me. Ursula played the controls of Wilson's instrument like a virtuoso, slowly changing the colors, moving them through a metamorphic sequence of quiet power. I was calmed and then awakened through the agency of pure color.

Only at the end did I notice the small footprint on the screen. During a therapy session for one of the children, Ursula explained, the child stood up and walked toward the field of color as one who has long been lost and suddenly sees the trail leading home. She caught his arm as his foot hit the screen. In order to grant the child's wish to live in color, Wilson designed and constructed an entirely novel therapeutic environment—a color therapy pool.

In a warm, dimly lit room, a tile wading pool is filled with water that shimmers and sparkles like a live jewel. As a therapist and child slowly make their way into the pool, I notice how brightly their bodies and limbs are illuminated. Wilson has installed strong colored lights around the perimeter of the pool just below the waterline. Very little of the light could escape because of total internal reflection, and as with my light box, without something off of which to reflect, light, as always, remains invisible. Only when the child enters is it clear that one is bathed in color as well as water. Wilson explained to me that for some children, it is only in the pool that they become aware of their own bodies. Swathed in color, they notice themselves for the first time.

Sunfield explored the therapeutic use of color with modesty and courage. Yet to use it effectively, they needed to know the inner character of color intimately and truly. Electromagnetic theory might help Wilson technically, but another kind of knowledge was required for its actual use. Here, Goethe's color studies were a natural point of departure for Wilson and the staff of Sunfield. For in them the inner aesthetic dimensions of color are as important as its outer aspects.

Color as the Character of Light

And so my nature studies rest on the pure base of living experience.[12]

Goethe

Goethe's scientific interest in light was born on a hillside outside Rome under the luminous blue dome of the Mediterranean sky.[13]

In conversations with landscape painters while he was in Italy, Goethe felt a significant disquieting dissatisfaction—they seemed to lack any real basis for the aesthetic use of color. What was it, he asked, that determined the artistic application of color? Was it only "caprice determined by taste which in turn was fixed by custom, prejudice, artistic convention"?[14] Conversely, were colors to be chosen in mimicry of external appearance alone, or was the artist guided by the cognoscenti whose own tastes defined the norm of the period's art? Goethe's temperament could not rest until he had seen through to the unitary principles of color, principles that would embrace also the "moral" dimensions of color. Once they were found, Goethe felt sure that these insights would have aesthetic implications. Significantly, Goethe's entry into color drew its initial impulse not from science, but from art.

After returning to his home in Weimar, Goethe opened a compendium of science to refresh his vague memory of the scientific explanation of color. He found there the conventional discussion of Newton's corpuscular theory, but it seemed of no value for his own ends. On the verge of abandoning the question, he reflected that he should be able to uncover the truth he sought about color purely through his own observations and experimentation. The realization led to his fateful request of optical equipment from Hofrat Büttner, and his first hallway experiment.

As Goethe lowered Büttner's prism from his eyes and turned away from the window, he also turned firmly away from the corpuscular theory of light advanced by Newton. Goethe was convinced that colors were not somehow secretly present in white light only to be sorted out by the prism. What Goethe saw convinced him this was an error. Darkness as well as light appeared to be equally necessary for the production of prismatic colors. Yet what Goethe really sought was not an alternative mechanical explanation for color production, but another way of explaining altogether. He rejected mechanical models of the kind common to his predecessors and contemporaries, but what were his alternatives?

Within Germany at this time, there flourished a philosophical movement that attempted to connect natural science to a lofty idealism. Friedrich Schelling, Lorenz Oken, and G. W. F. Hegel were but three of a much larger group of "nature philosophers" who would have been only too glad to have the giant Goethe join forces with them. Here, too, however, Goethe remained true to his own approach to light. The speculative philosophical thought systems of these countrymen were pale and unhealthy to Goethe's artistic eye. For example, Hegel, whom Goethe admired, and who was one of Goethe's few supporters in his battle with Newtonianism, offered up the following definition of light: "As the abstract self of matter, light is absolute levity, and as matter, it is infinite self-externality. It is this as pure manifestation and material ideality, however, in the self-externality of which it is simple and indivisible."[15]

If such were the fruits of Hegel's dialectical method, Goethe wanted nothing to do with them. Goethe felt compelled to make his own way toward light. Trusting to his artistic sensibilities, he sought out light where it lived, in the sensual phenomena of color. Here, not in Hegel's "oyster-like, grey, or quite black Absolute" (as Hegel himself called it in a letter to Goethe), could one read the life history of that precious resource called light.

Goethe declined to define light either mechanically or abstractly, saying simply that "In reality, any attempt to express the inner nature of a thing is fruitless. What we perceive are effects, and a complete record of these effects ought to encompass this inner nature. We labor in vain to describe a person's character, but when we draw together his actions, his deeds, a picture of his character will emerge."[16]

Attempt to define a human being in terms of psychological theories, and you inevitably fail to reveal his or her essential being, but as a novelist, describe how she walks, places her hand at her side, moves her head and mouth . . . and immediately the inner nature of a person is revealed. Goethe knew this method well. Therefore, he did not seek causes but a history of appearances in which light could show its manifold nature through color. These would combine to become a vivid and lively biography of light, whose text would lead one to an intimate and gentle knowledge of the being of light, spiritual as well as physical. As with humans, so, too, with our investigation of light, suggests Goethe. If we would know the nature of light, then we should look to its actions and gestures, which are colors.[17]

Colors, then, can act as the entryway to light's inner nature. He called his method a "gentle empiricism." Learning from Francis Bacon, but reaching beyond the narrow constraints of induction, Goethe practiced a style of investigation which, he wrote, "makes itself in the most intimate way identical with its object and thereby becomes actual theory. This heightening of the spiritual powers belongs, however, to a highly cultivated age."[18]

If we would anticipate that future, highly cultivated age and practice Goethe's "gentle empiricism" with light, then we must explore the full range of color phenomena offered by light, becoming "in the most intimate way identical with" them. By doing so, we move through color, becoming not only more intimate with the nature of light, but simultaneously initiating a transformation of self that leads to new faculties of insight.

—

WE TOUCH HERE on two aspects critical to Goethe's understanding of science. First, in common with all scientists, Goethe searched for patterns, the hidden lawfulness within the welter of color phenomena, but for him they were to be exalted perceptual experiences, not abstract substitutes for nature's glory.

Second, Goethe fully emphasized the significance of *Bildung* or "self-transformation" in his scientific methodology.

Fashioning Organs for Seeing

Goethe saw the human being as constantly engaged in a process of self-formation. We have learned that even natural organs such as the eye require imagination to see. If the blind cannot be given sight by physical means alone, then how much more true must this lesson be for those organs of cognition that allow us to "see" natural laws. The seeing of lawful patterns within the multiplicity of phenomena requires suitable inner organs. They are not given at birth, but develop in life. Nor should we confuse these capacities with analytic facility or logic, as valuable as these may be in their own right. In addition to analytical reasoning, all scientists (and we too) rely on a kind of seeing, a capacity for insight, that has been schooled through thoughtful experience. With it one sees that which others, staring at the same phenomena, may never see. In just this way, scientists make their observations and discoveries.

It has happened dozens of times. Every walk I take with an experienced naturalist convinces me that Goethe's attention to *Bildung* and the "organs" of insight was apt. Standing with a geologist before an outcropping of rock, he sees more than I who stand next to him. I make a few distinctions, he a hundred, and each one tells a story to him of which I know nothing: glaciation, a lake bed, or volcanic lava flow; he finds the fossil under my foot. I feel not only illiterate but blind. Not only does the geologist interpret the phenomena more fully, he sees things I miss utterly. I cannot even see the text, much less read it. He is Holmes, I poor Watson with all my bookish knowledge. As Emerson wrote, "We animate what we see, we only see what we animate."[19]

Such seeing can become refined, raised to a high art. As though through a powerful telescope, these cognitive organs bring the distant coastlines of uncharted scientific regions into focus. Through them, and not by analytical reasoning alone, are nature's essential patterns discerned and scientific discoveries made.

Remembering the "light of the body" can help us to understand Goethe's point. The sighted eye requires more than the input of natural light; it also requires Empedocles' inner, ocular light of intelligence. If we neglect the animating light of coherent intelligence that illumines and flows through all our senses, then the glory of the world stands mute before our inquiring spirit. Goethe emphasized the importance of a light that is within. In his words, "if the eye were not sun-like, how could we perceive the light?"

How does one kindle the Empedoclean fire of the eye? Goethe's reply was simply, by active engagement with the world. Our every thoughtful interaction is formative and profoundly educative: "Every object, well-contemplated, creates an organ of perception in us."[20] He considered the instrument of sight as a paradigm of organ development and formation. He understood it as crafted by light itself as it worked on the human organism. "From among the lesser ancillary organs of the animals, light has called forth one organ to become its like, and thus the eye is formed by the light and for the light so that the inner light may emerge to meet the outer light."[21]

Goethe's words harken back to ancient ideas, yet we can place them into a much more modern evolutionary context. Light is formative. Under its influence plants grow, but also the eye was formed. Similarly under the influence of mountains, stones, and streams, the geologist develops organs of cognition, kindles an Empedoclean light that elucidates his beloved kingdom.

Goethe's is a participatory science in which the sight of an idea, the epiphanous moment, is supreme. How can one come to see an idea within the phenomena, to see through color, the character of light? Only by building up the organs needed. The chemist laboring over her laboratory bench, the mathematician working at the blackboard, are working as much on themselves, educating capacities for insight, as they are on the chemicals in the flask or the equations on the board. Our every engagement and action is pedagogic, educating new organs for discovery, artistic or scientific.

The luminous eye slowly darkened during its long transit from Egypt to Alhazen and Descartes. In Goethe's hands it once again comes to light and life. Can we turn this new life back on light itself, hoping to penetrate to its essential nature? Can we develop the "eye" that will see light invisible?

—

THE HISTORY OF light is, in large part, a history of idolatry. One image of light after another has been offered in place of light. Goethe, like all the greatest scientists (Galileo and Newton included), knew the difference between the ingenious productions of the human mind and the laws of nature. Yet Goethe's position differed from theirs in his unswerving commitment to the concrete phenomenal dimensions of nature. His was an artistic, not a mathematical spirit, one that relished the smell of pigments and the sound of a well-turned phrase. If Goethe, like Faraday, would see into nature's eternal workings, then the facts or phenomena themselves would have to be raised to a high theoretical level so he could "see" in them the pattern nature was weaving. Like a portraitist working his pigments to reveal the character of the human being before him, Goethe worked with natural phenomena until they revealed to his attentive eye what he called nature's "open secret." His method was that of a consummate artist; but has a scientific discovery ever been made in any other way? Emerson rightly observed, "never did any science originate, but by a poetic perception." Goethe forcefully holds our attention to the epiphanous moments in science, to the poetry that is the heart of science.

Seeing Ideas

My perception is itself a thinking, and my thinking a perceiving.[22]

Goethe

Like Goethe reaching down to grab a prism from Büttner's case of optical equipment, we can begin to cultivate new eyes for light by putting a prism to our eyes to see the celebrated phenomenon of colors. Taking a prism into one's hand leads to a wealth of confused expectations. The unspoken question is, "I wonder what I'll see?" With our first look, wonder rises. Then, after a minute of smiles, a slightly intent, bemused expression of interest appears. It seems to ask, "How can I make sense of this lovely jumble?" In that moment, we step away from the pure pleasures of naive "empirical phenomena," as Goethe called them, and toward the arena of "scientific phenomena."[23] We move from wonder, through interest, toward insight, and so approach the sight of the ideal.

It happens slowly, with the experimenter beginning to change the conditions under which the phenomenon occurs, distinguishing those that are essential to the effect from others that are not. One is quickly led to simplification. Both light and dark are needed to produce prismatic colors, and the simplest arrangement is a single straight boundary between light and dark regions.

The illustration on page 208 shows the two significant configurations, light above, dark below; or the opposite, light below and dark above. When viewed through a prism, blues and violets appear along one edge (which edge depends on how you look

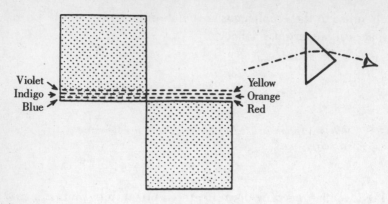

When viewed through a prism, edge colors appear where light meets dark.

through the prism); along the other edge are yellows and reds. These two arrangements of light and dark are complements of each other. When seen through a prism, they yield the "cool" colors for one arrangement and the "warm" colors for the other. To Goethe's eye, the very simplicity of the experiment and the character of its results supported its importance. Recall also the aesthetic question, still in Goethe's mind from his Italian journey. The polarity of warm and cool in color was a well-known characterization among artists. Here it occurs again quite naturally, but in a scientific setting. One arrangement of light and dark yields the warm colors, its opposite the cool. Color's artistic polarity seems to have an objective basis in experiments as well.

Only green and magenta are missing if we wish to complete the full color circle. These magically appear by modifying the edge arrangements to form thin bands of light and dark, green appearing in the middle of the white band, and magenta in the black band.

In order to understand the foregoing phenomena, Goethe did not seek an abstract theory, but sought special examples that best represented those relationships of light to darkness which

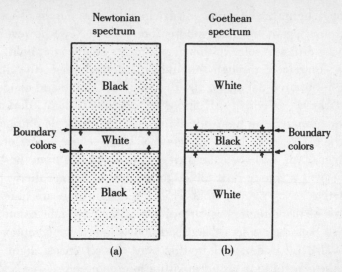

Green appears as the white region narrows (a). By contrast, as the black region narrows, magenta appears (b).

are responsible for the appearance of warm and cool colors. These he called *"Ur*-phenomena," that is, primal or archetypal phenomena. In the case of warm colors, the yellows and reds of dawn and dusk offer us the instance we seek. By learning to see them, we come to comprehend one pole of the mystery of color.

As the sun sets, its light passes through more and more of the atmosphere en route to the eye. Thus, in its journey from sun to us, light passes through the darkening or "turbid" medium of air, as Goethe called it. In the process, all the warm colors arise. This is the archetypal relationship between light and darkness that yields red, orange, and yellow—light through darkness. The stronger the darkening, the redder the color.

The blue vault of the daytime sky offers us the archetypal instance of the other pole of color. Here light does not pass through darkness, but just the opposite—darkness passes through light. Looking up, we gaze into the dark depths of space, but once again the atmosphere intervenes. Now, however, it plays a different role. In this case, the air catches the light. We look, therefore, through the light-filled medium of the atmosphere into the darkness. Or, if we follow Goethe and conceive of darkness as equally, if oppositely, active to light, darkness shines through the light-filled air, and the cool colors arise.

Once you learn to see the law of color in the colors of the heavens, you will see examples of it everywhere, from the blue haze over a smoky pool table, to the use of "atmospheric perspective" by a painter (that which is distant appears blue because of the intervening turbidity of air). Prismatic colors, as in the boundary colors discussed above, are more complex but can also be understood in this way. In all cases, light and darkness meet in a turbid medium to create color.

The account of color production given by the most recent theories of physics offers a similar, if far more exact and mathematical account. In them, colors arise through the "scattering" of light. The turbid medium provides innumerable scattering centers, be they molecules in air or a glass prism. From them light is scattered according to strictly mathematical laws, and in the process colors are produced. Even the rainbow, set between Alexander's dark band and a luminous interior region, can be understood in an analogous way. Where light meets darkness, colors flash into existence. Colors are, therefore, the offspring of the greatest polarity our universe can offer. In the mythic language of Zarathustra, colors are a reflection of the mighty battle relentlessly waged between the god of light, Ahura Mazda, and the dark hosts of Ahriman. In Goethe's language, "Colors are the deeds and sufferings of light," the deeds and sufferings of light with darkness.[24]

If we follow Goethe's pathway into color, we are not led to models of light in terms of waves or particles, but to a perception of those relationships between light and darkness that give rise to color. Seen aright, the phenomena of the blue sky and sunset are the theory, and true to its Greek root, *theoria*, theory really is a "beholding." In Goethe's words, "The highest thing would be to comprehend that everything factual is already theory. The blue of the heavens reveals to us the fundamental law of chromatics. One should only not see anything further behind the phenomena: they themselves are the theory."[25]

Like the geologist reading rocks, or Newton seeing the apple fall, or Archimedes crying "Eureka!" we can grow to perceive the laws of chromatics in the blue of the heavens and the first light of dawn. Through studying the action of light in darkness, and darkness in light, we come to sense the "deeds and suffering" that are color. Once we have kindled an Empedoclean light within, fashioned the requisite organs of insight, the archetypal phenomena appear, and in them we see an idea.

—

PHILOSOPHY SINCE PLATO has divided knowing into two insular realms: ideas and experience. Naively, but tenaciously, Goethe ceaselessly sought a way to experience ideas, to bridge the chasm others thought unbridgeable. The union of idea and experience may seem impossible, but as Goethe says, "nothing forbids us from seeking a loving approach to that which lies beyond our reach."[26]

Goethe's method gradually converts facts to theory, seen reality to ideal reality, and is Goethe's response to philosophical dualism. One cannot take truth by force, but perhaps indirectly, through phenomena, sign, and symbol we may approach her. "The True, which is identical with the divine, does not allow itself to be recognized by us directly. Rather we discern it only in reflection, in instance, symbol, in particular and kindred appearances. We become aware of it as incomprehensible life and yet cannot renounce the wish to comprehend it."

Goethe's method requires a reciprocal enhancement of both

natural phenomena and the observing mind. It all begins with wonder, as Plato rightly said, but then passes on to interest and so to active inquiry. In the process, new organs of perception are fashioned that are suited to seeing the essential aspects of the phenomena before us. As we enhance our cognitive capacities we simultaneously enhance the world we see until, ultimately, we behold the ideal within the real as archetypal phenomenon. To them one rises, as Goethe described the process, and from them one can descend in order to understand specific phenomena. They are the ultimate experience, and the limit beyond which one cannot legitimately go. Most of us, however, do not recognize this and so go on, putting, for example, a model or idol in place of the archetype. "The sight of an archetypal phenomenon is generally not enough for people; they think they must go still further; and are thus like children who after peeping into a mirror turn it round directly to see what is on the other side."[27]

—

GOETHE'S SENSE OF scientific understanding is grounded in insight, not model building, and so is true to the heart of both science and art.

Every scientific discovery from Galileo to Einstein can trace its origin to the eureka experience in which a phenomenon becomes transparent to the ideal, and an idea is seen. From this exhilarating moment, the scientist works to translate his or her insight into words and symbols. In the process, the eureka experience is often lost while its technical power is retained. Goethe was more interested in the former, seeking constantly for means that would permit everyone to have their own epiphany into nature's ways, to see ideas.

—

GOETHE PERFORMED FOR philosophy a common piece of parlor magic. The magician stands with two disconnected solid metal rings before the audience, one in his left hand, the other in his right. He taps them together to prove the impossibility of their union. Then, before their very eyes, he clangs the two and they are linked. Each ring passes through the middle of the other. In an instant the topology has changed utterly. Now all such distinctions as inside-outside simply lose their meaning. Like the two rings, Goethe considered the realms of thinking and perceiving as interpenetrating. Perceiving is at once outside and at the center of thinking, and thinking likewise passes through the heart of seeing, and surrounds it.

Two worlds, kept so long apart, are united in our perception of archetypal phenomena. To see them we must fashion new organs of cognition, for they cannot be gained by logic alone. Once known, they represent the highest we can hope to attain. To the artist, one final and all-significant aspect of this method is that in perceiving the archetypal phenomenon, one does not denude or degrade nature but exalt her. The sunset is still gloriously red, not reduced to differential absorption and scattering. The perception of a scientific idea does not require the death of the beautiful.

The last chapter of Goethe's *Theory of Color* undertakes a preliminary treatment of the "sensory-moral" effects of color. In these pages, Goethe describes his inner response to color, connecting it back to earlier sections of the book. The eye's tendency to complete the color circle is related to the principles of color harmony; the polarity of warm and cool colors takes on new meaning in light of his prism experiments. The inner aspects are as much a part of the experience of color as the redness of red. The archetypal phenomena include the moral with the sensual.

Remember Goethe's motivating question framed on a hillside outside Rome regarding the use of color by artists. Remember, too, the therapeutic use of color at Sunfield Children's Home. Textbook physics cannot answer the real questions of Goethe or Wilson. Such answers come only by working with the phenomena themselves, because then we fashion organs for a science in which the beautiful as well as the useful, the human as well as the physical, can be experienced. Thus, Goethe performed a second piece of conjuring. As Gaston Bachelard once wrote, and as Goethe amply demonstrated, "The phenomena of the world, as soon as they acquire a little consistency and unity, turn into the human truths."[28]

More Light!

Throughout his life Goethe was a lover of nature and especially of light. When he was young, that love was passionate; when he was old, it became quiet but intense. As a child, Goethe revered God through his works, his creation of minerals, plants, animals, and heavens. Above all these ranked the sun. To this god, the child Goethe once constructed an altar after the fashion of Old Testament prophets. On his father's ornate, red, four-sided music stand he arranged his most precious specimens: crystals, ores, shells, and plants. Yet something especially fine was needed for the summit—a flame with gently rising smoke, perhaps. In a small porcelain saucer the young priest placed a tablet of incense, completing the altar. Lighting the tablet was all that remained in order to consummate the ceremony.

The secret service occurred at dawn with only Goethe in attendance. As a brilliant sun rose above the apartments to the east, Goethe used a magnifying glass to focus the sunlight onto the incense. Like a Zoroastrian priest, the child connected the sacred fire atop his altar to the sun. The liturgy was complete. Through the power of the sun, and with a little technical assistance from the lens, the mystery was enacted. Johann was content.

Goethe lived long enough to see his color science ignored or rejected by scientific contemporaries in whom mechanical conceptions of light were unalterably rooted. They preferred a mathematical language to that of color experience, physical models to archetypal phenomena. Yet this greatest of Germany's many geniuses never swerved from his own judgment concerning the high value of what he had accomplished.

The light of the ferryman's lamp in Goethe's fairy tale *The Lily and the Green Snake* turned all it touched to gold. Similarly, colors are the precious ripples that sparkle in light's wake. For decades Goethe studied them, saying at his life's end, "I have known light in its purity and truth, and I consider it my duty to strive after it."[29]

In his final conflict with death, light was still Goethe's last request. A half hour before his passing, Goethe commanded that the window shutters be opened so that more light might stream into the room where he lay. His earthly striving at an end, how fitting that Goethe's last words are said to have been: "More light!"[30] Ruskin was right, those who love color are pure. Surely to them, if to anyone, will be granted nature's open secret—light.

—

AT SUNFIELD I came to appreciate Goethe's approach to color, and his understanding of the essence of scientific inquiry as a seeing of ideas, but his treatment of the "sensory-moral" aspects of color, so important to Wilson's color therapy, still puzzled me. Orthodox physics had no room in it for moral dimensions of color. I asked Wilson how he understood them. In response, he told me a story. When he was a young man, training on the violin and for orchestral conducting, he worked with an opera company. During this time, he became fascinated by stage lighting and, through that, by color. His mother gave him, therefore, what was at first a highly puzzling book on color by a thinker she admired. After reading half of it, Wilson threw the volume across the room in despair, only to pick it up again later. In the intervening years, he had become a careful student of this same philosopher and suggested that if I wished to know more about the moral and spiritual dimensions of color as Goethe conceived them, then I had best read him, too.

Rudolf Steiner's Metaphysics of Light

Only a weak light glimmers, like a tiny point in an enormous circle of blackness. This weak light is no more than an intimation which the soul scarcely has the courage to perceive, doubtful whether the light might not itself be a dream, and the circle of blackness, reality.

Vasily Kandinsky

While in Berlin during 1908, the painters Vasily Kandinsky and Gabriele Münter met old friends from their Munich days, Alexander and Maria Strakosch. Alexander Strakosch recalled how they, together with other dear friends from Munich, would meet, spending wonderful evenings together in Kandinsky's Berlin atelier. It was a time of profound seeking, personally and artistically, for them all. What they sought could not be provided by the mechanical imagination of nature and self offered by nineteenth-century science. Many felt the need for a more expansive and ensouled imagination of both. Strakosch wrote, "Many friends, knowing of our searches, drew our attention to the weekly lectures at the *Architektenhaus* as essential to hear."[31] These were the lectures of Rudolf Steiner, the same scholar and philosopher who, in the early 1890s, had been in Weimar as editor of Goethe's scientific writings. He had left the stodgy atmosphere of the Goethe-Schiller archives for the fervor of Berlin, where, in 1897, he assumed editorship of the avantgarde *Magazine for Literature*, a position that placed him squarely in the lively literary circles of Germany's greatest city.

The audience at the *Architektenhaus*, however, had not gathered for literary ends, but rather to hear Steiner speak about his spiritual philosophy of man and universe. By the time Kandinsky and friends heard him in 1908, Steiner's activities in Berlin had changed fundamentally from those of a Berlin intellectual to a spiritual philosopher. He had already written and lectured extensively on meditation, Christianity, the spiritual history of mankind, and the spiritual dimensions of our universe, mostly to Theosophical audiences (he would found his own Anthroposophical Society in 1913).

On March 26, 1908, Kandinsky, the Strakosches, and their company attended Rudolf Steiner's *Architektenhaus* lecture on "Sun, Moon and Stars." The content of the lecture clearly placed Steiner with those who understood light in terms of spirit, that is, with figures such as Grosseteste, Mani, and Zoroaster. He, like Grosseteste, stood with a foot in two worlds, one in the world of spirit and the other planted firmly in the world of science. Here was the essential and eternal tension out of which Steiner sought to create harmony instead of discord.

The lecture must have struck a special chord in Kandinsky, for directly after it he painted the Ariel scene from Goethe's *Faust*, lines of which Steiner had read at the close of the lecture. Kandinsky gave the painting to Maria Strakosch. In it a healing rainbow hovers above a waterfall. Suspended mist calmly rises from a watery tumult, catching the light, bending it into the panoply of colors that bathe a broken Faust. Life is like the rainbow. We have heard the lines before:

> *The rainbow mirrors human aims and action.*
> *Think, and more clearly wilt thou grasp it, seeing*
> *Life is but light in many-hued reflection.*

Light is connected by Goethe to life; the rainbow reflects not only sunlight, but human aims and action. Its colors are an image of our aspirations and deeds. Here was the seed for an imagination of light that did not rob artistic imagination as it enriched scientific understanding. Kandinsky and Steiner understood Goethe's metaphor and each gave it expression in his own medium.

—

THE SON OF a minor railway official, Rudolf Steiner had been educated at the Vienna Technical University, the MIT of Austria, completing an undergraduate degree in mathematics, physics, and chemistry. In addition to the usual diet of technical subjects, he closely studied the works of Kant, Fichte, Hegel, Nietzsche, and Darwin, later earning his Ph.D. in philosophy while editing Goethe's scientific works in Weimar.[32] Steiner knew science and academic philosophy, but from his personal experience of spiritual phenomena, he felt that a broader conception of the world was required than that offered by nineteenth-century thought. Early on, the treatment of light became a paradigmatic instance of the deep conflict he felt between his scientific training and personal spiritual experiences.

As a university student, Steiner, like others in his day, grew critical of the wave theory's ether hypothesis.[33] Yet Steiner went much further. He felt that while light could manifest as color in sense phenomena, light itself was essentially *extrasensory*. In it one met the purely spiritual within the sense-perceptible. As he wrote, "light is already spiritual. Within the sense-perceptible the spiritual holds sway."[34]

Not only light, but much of existence needed to be understood spiritually, as well as physically, in Steiner's view. Orthodox nineteenth-century science was ill equipped and undisposed to take up this task, so on the basis of his philosophical studies and Goethe's scientific investigations into color, botany, and biology, Steiner proposed a "science of the supersensible." His *Architektenhaus* lectures were but a single instance of that ambitious project.

As a complement to the justifiable physical investigations of light, Steiner offered a spiritual imagination of light. As we have seen, ancient civilizations invariably viewed light and its heavenly sources as divine. Steiner sought a modern Christian metaphysics in which the cosmos celebrated by Egyptian priests, Greek philosophers, and Robert Grosseteste could find new life on a philosophical footing at peace with science. As long as the observations, experiments, and thinking of natural science remained strictly objective, Steiner felt that no conflict could arise with spiritual science. Often, however, natural science extends its reach beyond its observations, refusing a place for the spirit. Such a position was, to Steiner, philosophically untenable and morally bankrupt. He felt the time was ripe for a spiritual science that could move from sense phenomena to spiritual phenomena without lapsing into vague mystical utterings. He rejected the then-fashionable approach of spiritualism and psychic research as simply a displaced materialism of the spirit, and sought rather to follow Goethe's scientific methodology, which implied the development of organs of cognition suited to every domain of experience, including domains of the spirit.

Spirit Light

The universe seen from within is light; seen from without, by spiritual perception, it is thought.[35]

Rudolf Steiner

Like so many before him, Steiner drew on the opening passage of the Gospel of St. John when seeking the spiritual dimensions of light. John speaks there of the Logos—or Word—as the Light of the World. Everything we see, from sun to stream, cloud, animal, and man is, says Steiner, an embodiment or image of that divine-spiritual reality—the Logos.

The purest appearance of the outer physical body of the Logos is the light of the Sun. Sunlight is not only material light. To spiritual perception it is also the garment of the Logos. In sunlight, spirit streams to the earth, the spirit of love. . . . Together with physical sunlight streams the warm love of the Godhead for the earth.[36]

Light is the pure body of the Word, of the Logos. The very same light that illumines our world is the most perfect image of God's creative song. Grosseteste had called light the first corporeal form; Steiner agreed. The electromagnetic theory of light was, in Steiner's view, but a pale and abstract reflection of light's much greater nature, one of which we should remain forever mindful.

—

IN LECTURES AT the close of 1920, given to an audience well acquainted with his spiritual world conception, Steiner offered a remarkable vista onto the nature and biography of light. He performed a kind of spiritual archaeology of the ancient origins of light, providing an account that was mythic in its proportions and language, yet one he intended to be an exact imagination of the genesis and being of light.[37] Once again Goethe provides the entry point for our considerations. "Colors are," Goethe said, "the deeds and sufferings of light." The metaphor is exact; yet what can act and suffer but beings?

The spiritual hierarchies spoken of in early Christianity (Angels, Archangels, and so on), and which were so much a part of Grosseteste's world, were and still are real according to Steiner. To each of us is given a personal guardian Angel, over groups there hovers an inspiring Archangel, and a *Zeitgeist* or Archai rules every age. Steiner's many descriptions of the nature and evolution of the hierarchies were detailed and of a scope unknown since the Gothic cathedral builders. It seems that the angelic orders have evolved over enormous periods of time in ways somewhat similar to man's own spiritual evolution. In particular, in aeons past, the angelic hierarchies passed through a "human" stage. During that period they possessed an inner or moral world analogous to ours, one full of life and struggle, of deeds, noble and ignoble. Then, in a remarkable passage, he suggests that what the angelic host once carried within themselves as moral realities have since become "world-thoughts," which we in turn experience as the light of the world. The actual physical light that surrounds us is the fossillike remains of an ancient moral world lived by angelic beings. What was inner becomes outer.

With disarming logic, Steiner goes on to say that the moral world we are now nurturing within our souls will likewise one day become the light, or darkness, of a future evolutionary phase of the cosmos. "We see about us today a world of light; millions of years ago it was a *moral world*. We bear within us a moral world, which, millions of years hence, will be a world of light. . . . And a great feeling of responsibility toward the world-to-be wells up in us, because our moral impulses will later become shining worlds."[38]

Within such a view, it can never be a question of separating the moral from the physical. In Emerson's words, "Every natural fact is a symbol of some spiritual fact. . . . The world is emblematic. The laws of moral nature answer to those of matter as face to face in a glass."[39] Thus, every physical reality is but the obverse of a moral reality. Steiner's vision extended Emerson's to the greatest limits of human and cosmic history. In his view, the physical world is the fruit of the moral world. Pure hearts truly will illumine future worlds. Or, if we harbor darkness within, then a dark world will be its lawful consequence in the distant future. We are cocreators of the world not only through the deeds of our hands but, in even greater measure, through the spiritual impulses we foster inwardly. In Steiner's words,

"The physical and the moral do not exist side by side. No, they are only different aspects of something that is in itself one. The moral world order reveals itself out of the natural world order."

The connection of physical and moral, of sensual and spiritual, was, and remains, largely heretical within religious as well as scientific circles. Powerful forces within Protestant theology have insisted on dividing religion from science utterly. Likewise, within the scientific community, religion has been viewed as treating an aspect of life entirely distinct from those investigated by science. Max Planck spoke for many when he said, "There can never be any real opposition between religion and science; because the one is the complement of the other." For scientist and theologian alike, there have been two worlds, and so two truths: one scientific and the other religious.

Goethe, Emerson, and Steiner, by contrast, acknowledge distinctions within life, but refuse to fragment it. Truth might be many-sided, but at root it is one. To Steiner, Planck's views are naive. Simple faith, as Planck envisions it, cannot withstand the impressive and ever-growing force of natural science. If modern habits of visualizing nature as a deterministic mechanism prevail, "then there is no possible way of saving the moral realm . . . there is simply no evidence anywhere in this moral realm of a power to prevail against the realm of the natural order."[40] By implication, if the moral realm falls, so, too, does the light of future aeons. Thus, Steiner sees the segregation of scientific and spiritual knowing not only as a hindrance to a comprehensive understanding of light, but also as a threat to our future. To Steiner, the natural world around us grows out of the moral world within us just as the butterfly evolves from the caterpillar. Its shape will depend on the worm we place inside the chrysalis.

—

GOETHE'S COLOR THEORY had been either neglected or dismissed. Steiner's conception of light, together with his entire philosophy, met with more violent reactions. Attacks on him mounted during the early twenties from all points of the compass: clergy, academics, scientists, politicians, and proto-Nazi groups, to name a few. Most occurred in the press, but others were more brutal. During a 1922 public lecture in Munich, Steiner escaped serious harm only through the quick action of a cadre of youth who placed themselves on the stage between a group of assailants and the speaker. The touring agency of Sachs and Wolff informed Steiner that they could no longer guarantee his safety, and therefore would not arrange further public lectures for him.

On New Year's Eve, 1922–23, following a lecture by Steiner on the history of natural science, the headquarters of his Anthroposophical Society located in Dornach, Switzerland, was set afire by an arsonist. The structure, which seated over a thousand, had been named the Goetheanum after Goethe. Lovingly crafted, carved, and artistically painted by hundreds of volunteers over the previous ten years, it was the impressive flower of Steiner's Anthroposophy. In that single night, Rudolf Steiner and his collaborators watched as their years of labor were lost in flames, in the moral darkness of persecution.

Broken but not defeated, Steiner redoubled his efforts during the final two years of his life. He formed a University for Spiritual Science and designed a new organic, concrete structure to become the second Goetheanum. It became the home for the impulse to which he had given voice, one that affirmed the union of religion, science, and the arts, and which saw the moral world of man as the true progenitor of light.

—

WITH GOETHE AND Steiner, a major contrapuntal theme is introduced into our biography of light. Until now the story line has flown smoothly from ancient mythic to modern scientific imaginations of light. These two figures, however, mark a renaissance of the mythic. Both were citizens of the modern age, both understood science, and yet they sought to transform it so as to include within it light's spiritual dimensions.

To my eye, our world needs the enrichment of myth and an ethic born of compassion. Every laboratory should be sited as Wilson's was, on a sunny field, where scientists remain cognizant of the questions suffering poses. In the figures of Goethe and Steiner, I sense the germ of a new mythology and of a science of compassion.

Novalis wrote wisely when he said that within the flame, all natural forces are active. The truth of his words rests on the fact that the flame, and the light it sheds, are as much moral and spiritual forces as natural ones. Light is not two things, but one. Its efficacy lies in the unity. The scientific study of light never need diminish its full stature. But we, like Navajo gods, walk a fragmented landscape riven by canyons and gorges. Unlike them, we are timid, holding fast to the firm familiar soil beneath our feet, unwilling to throw a rainbow bridge across the dark chasms of our world. Not lunacy, but courage is needed to see our world whole, to know that love must become concentric with insight.

CHAPTER 9

Quantum Theory by Candlelight

Some candle clear burns bright somewhere
  I come by.
I muse at how its being puts blissful back
With yellowy moisture mild night's blear-all
  black,
Or to-fro tender trambeams truckle at the
  eye.

Gerard Manley Hopkins,
"The Candle Indoors"

On the evening of October 21, 1929, America relived one of the great moments in the biography of light. That night, Thomas Alva Edison, Henry Ford, and President Hoover strode together down the main street of Ford's new living history park, Greenfield Village, to Edison's old laboratory (moved there by Ford from Menlo Park, New Jersey). At his old lab bench, frail with age and overcome with emotion, Edison sat ready to reenact the final moments in the creation of his incandescent electric light of fifty years before. Broadcast over hundreds of new radio stations to millions of listeners, Ford asked everyone to turn off all their electric lights in honor of the occasion. That evening, all of Greenfield Village, and much of America, awaited Edison's invention by candlelight. When the glorious moment came, Edison threw the switch that fed electrical power to his bulb's fragile carbonized filament. As its feeble glow lit his aged face, in unison, America threw the switches that fed power through the circuits of Greenfield Village and darkened homes across the continent. For a few minutes, the light of all past ages had been called back to life so that it might be ritually extinguished in a triumphant moment of technical bravado.

Now, even more than in 1929, the candle is an anachronism, but perhaps as a consequence it has taken on a sacred connotation. We, like many before us, place its small fire upon our altars, by our bedside, or, if we have the courage, on the boughs of our Christmas trees. We look, and as Gerard Manley Hopkins put it, we "muse at how its being puts blissful back / With yellowy moisture mild night's blear-all black." Yet the candle flame seems to give off more than light. In a time of utility, it is more a symbol than a household technology. And like the rainbow, it, too, has been a doorway into the understanding of light.

—

EACH CHRISTMAS FOR thirty-five years Michael Faraday gave a series of lectures "before a juvenile auditory," which quickly became immensely popular. During the holidays of Christmas 1860–61, he gave his last series of juvenile lectures in the great theater of the Royal Institution on *The Chemical History of a Candle,*[1] lectures that have since become classics in the history of science. Although seventy years old, he spoke with the voice of youth. "I claim the privilege of speaking to juveniles as a juvenile myself," he said; and so he did, with a joy and zest that belied the fifty years of service he had rendered the Institution. The apprentice bookbinder had become England's most distinguished and beloved scientist. Faraday invited his young listeners to join him in his investigations, and as he told them, "There is no better, there is no more open door by which you can enter the study of natural philosophy than by considering the physical phenomena of a candle."

Light a candle and notice first the perfect cup formed below the flame to carry the melted wax. The flame, reaching down the wick, melts the wax at the candle's center, while a current of air rising around the candle keeps the rim cool and high, and so creates a vessel perfectly suited to hold its molten contents. The liquid within is drawn up the wick by the same forces that draw sap up a tree or plant: capillary action. Instead of feeding leaves and flowers, however, the liquid wax vaporizes in the dark inner region of the flame closest to the wick, mingles there with the air, and feeds the flame. If this were all, as it is for some flames, a candle would shed little light. The bright yellow cone that spreads its gentle radiance, however, is due to tiny glowing embers of unburned carbon, the same that turn up as soot when the wick is too long. Cold, it is the blackest of substances, but when hot, soot becomes beautifully luminous.

To the poetic eye of Gaston Bachelard, the candle flame is a model phenomenon. In it "The most vulgar material of all produces light. It purifies itself in the very act of giving off light. Evil is the nourisher of good. In the flame, the philosopher encounters a *model-phenomenon,* a cosmic phenomenon, a model of humanization."[2] As model phenomenon, the candle flame is symbolic, incorporating into its nature a moral as well as a physical aspect. Coarse matter is purified in the flame becoming light. Looking at a flame, Paul Claudel wonders at the transformation it effects: "Whence matter takes flight to wing itself into the category of the divine."[3] To the poet, it offers a model of humanization, to the scientist, an unsolved riddle; either way, the candle flame draws us toward it like moths.

In his Christmas lectures on the candle, Faraday neglected to show his audience of children one feature which, beyond all others, held within it the germ that revolutionized physics and ushered in a fundamentally new conception of our physical world. That neglected feature was the color of candlelight.

Look at a candle flame. It flickers, waves, and pulses, a luminous, yellow, insubstantial form. View it now through a prism and there appears the familiar sight of the spectrum as a beautiful, flame-shaped array of rainbow colors. Within the delicate colorful form lies hidden an entirely new understanding of nature. "In the flame of a lamp all natural forces are active,"[4] wrote Novalis. Its peaceful flame offers, as Faraday has told us, the surest doorway to knowledge. We must pass through it to make our way to the quantum theory of light.

Light from Heat

Light is the genius of the fire process. Light makes fire.
Novalis

Any heated solid or liquid, whether it be clay in a potter's kiln or the filament of an incandescent lamp, glows in a way entirely similar to soot in a candle flame. It is a universal law that is oblivious to the substance being heated. Hot matter always gives off light, and always in the same way.

Every potter, glassblower, and metalworker knows this law. The colors of the heated, radiant substances with which they work change predictably with temperature. At low temperatures the blacksmith's iron is the same dull red as the potter's clay. The hotter either becomes, the brighter and more orange and then yellow the glow of the material. Every temperature has a unique color (the so-called color temperature). This is the basis for optical methods of temperature measurement often used in glass and metal shops. In astronomy it allows scientists to determine the temperature of the sun's surface to be eleven thousand degrees Fahrenheit. Heat and the color of light are linked.

—

ANYONE INTERESTED IN the light of a candle or that of the sun might naturally send its light into a prism in order to examine the spectrum produced. Just such experiments were being done in England and Germany in the nineteenth century with startling results. While the English astronomer Herschel was examining a solar spectrum, he caused it to fall on a row of thermometers. He was not the first to do so, but he noticed something that others had not. The thermometers that accidentally extended *beyond* the visible red edge of the spectrum registered a significant rise in temperature. Where nothing was to be seen, a *heat* effect was still produced. Similar observations were being made by others about the same time but at the other end of the spectrum, beyond the violet edge. Now, however, the effects produced were not heat but *chemical* in nature. Minerals changed color or glowed when placed beyond the violet end of the spectrum. These investigators had discovered what we now call infrared and ultraviolet radiation. The radiant heat one feels from a cast-iron stove is an example of invisible infrared radiation. At the other end of the spectrum, not only minerals change color in ultraviolet radiation, but so do we. Our suntan depends in large part on ultraviolet light, as does photosynthesis in plants.

These forms of radiation were found to behave like visible light in all respects except that they produced no response in the human eye. Like dogs that can hear pitches outside the range of human hearing, creatures do exist that evidently can

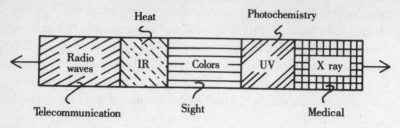

The electromagnetic spectrum.

see in these parts of the spectrum. We, on the other hand, must rely on less direct means to demonstrate their existence. The discovery of "invisible forms of light" broadened enormously the conception of light held by physics. Nowadays we conceive of it as running the entire gamut from radio waves to gamma rays. The spectrum of candlelight is far broader than meets the eye.

Many experimental studies of light from glowing hot bodies were being done in Berlin in the last years of the nineteenth century. The complexity of the candle flame with its varied burning conditions and temperature regions led the Berlin scientists Rubens, Lummer, and Pringsheim to invent a light source for which the temperature-color relationship could be studied more minutely. As a consequence, theirs became the nineteenth century's most accurate spectral analysis of light emitted from hot bodies. For technical reasons, it was called the study of "black-body radiation."[5] Conceptually the experiment simply quantified what was obvious to the eye when it viewed a glowing body through a prism, but took into careful account the invisible, infrared part of the spectrum.

At a given temperature a very particular, continuous spectrum of colors is seen. For example, when, to the eye alone, hot metal looks orange, then the brightest color in the spectrum is orange with colors to either side (red to one side, and yellow-green-blue to the other) fading off gradually in intensity. The specific form of the intensity distribution, color by color, was the function carefully measured by Rubens and others in Berlin. The question was, how to understand it, that is, how to understand the colors of candlelight?

As we have seen, the nineteenth-century physics community was confident that between the mechanics of Newton, the electromagnetic theory of Maxwell, and the pristine science of thermodynamics, it could account for everything. One needed only to increase the computational power available and all unresolved problems could be solved. Certainly the modest light of Faraday's pretty candle was child's play, and should be easy game for the collective scientific wisdom of the previous three centuries. And so the best theoretical physicists of the period in both Germany and England set out to calculate from first principles the spectrum of a candle flame. Yet they failed. In 1899, physics could not account for the colors of candlelight!

An Upright Man

The study of light has resulted in achievements of insight, imagination and ingenuity unsurpassed in any field of mental activity; it illustrates, too, better than any other branch of physics, the Vicissitudes of theories.

Sir J. J. Thomson, 1925

During the closing years of the nineteenth century, an industrious theoretical physicist had joined the faculty at the University of Berlin. His name, now inseparable from the birth of quantum mechanics, was Max Planck.[6] The descendant of a line of pastors, scholars, and jurists, Planck took from them the conservative and upright bearing that he maintained throughout his life. His ancestors had been exemplars of the traditional

Enlightenment values of rationalism and tolerance, joined with strong ecumenical Protestant convictions. Max Planck followed their good example, standing in the eyes of his contemporaries as a paragon of classical moral precepts, as well as master of all the best fruits of classical physics. Yet life is complex. His Prussian loyalties would later demand unsavory compromises with Nazi science, but in 1900 his tenacious intelligence was forcing him reluctantly to cross the border to a new imagination of light.

At the instigation of a friend, Planck embarked upon his theoretical analysis of light, fully expecting that classical physics would be adequate to the task. Over the course of time, however, try as he might, none of his efforts bore satisfactory fruit. As did others, Planck began by assuming that luminous bodies could be modeled as a collection of atomic oscillators vibrating at distinct frequencies. With egalitarian impartiality, the laws of thermodynamics distributed the available heat energy evenly among them. All calculations based on this model, including Planck's, were in catastrophic disagreement with the data. In 1899, after four long years of faltering, Planck made a seemingly small but very significant concession. In his theory of black-body radiation, he condescended to an apparently unjustified mathematical artifice whose full import he dimly sensed but did not then, or for several years thereafter, fully appreciate.

Consider the swinging mass of a pendulum clock. Its frequency of oscillation is determined, as Galileo first discovered, by the length of the pendulum. Its amplitude (that is, how far it swings) is determined only by how it is set into motion, by the energy imparted to the pendulum. Strange as it seems, Planck was forced into making the extraordinary assumption that at the quantum level, this description of the pendulum's motion is inadequate. In truth, not all amplitudes are allowed. That is, we cannot pull the pendulum back to any height we wish, but only to certain initial heights. Thus, the energy of the pendulum cannot take on all values. Quite to the contrary, only a discrete set of such amplitudes and energies is possible. They are specified by Planck's famous relationship: $E = nh\nu$. That is, the energy of an oscillator, such as a pendulum, can only take on values that are integer multiples of h (Planck's constant, a *very* small number) times the frequency ν (the number of ticks per second).

This single assumption changes everything. Energy can be given to or taken from the pendulum only in specific increments. Like walking up stairs, you cannot take half a step at a time. *And* for high-frequency oscillators (corresponding to blue light), the steps are bigger than for low frequencies (red light). In addition, at low temperatures not enough energy is available to excite the high-frequency blue oscillators or, put another way, to take the large steps needed. But other oscillators are around with smaller "red" steps that can be more easily climbed. Planck's analysis, therefore, led to the prediction that at low temperatures, the red oscillators alone would be active in absorbing and emitting energy. Consequently, the spectrum of candlelight (low temperature) would be predominantly red, and the missing blue part of the spectrum would appear only for higher temperature phenomena.

At root, the energy given to or emitted by the oscillators is electromagnetic, that is, light. Thus, Planck's assumption, if taken literally, implied that light itself might be quantized, existing only in discrete units. Planck initially thought his assumption was merely a mathematical trick that expedited his calculation, and would one day disappear. The English physicist James Jeans agreed, suggesting that h be sent to zero at some point in the calculation, and all would be well again with the world. Yet Planck's constant refused to be dismissed. Pandora's box of quantum mechanics had been opened, and all the attendant ills that flowed from Planck's analysis could not be stuffed back into the tidy enclosure of nineteenth-century physics. Herschel's "box of light" was in apparent disarray. Out of this dilemma was fashioned the new quantum of light. The colossal implications of Planck's modest assumption were apparent only to a few of his peers. Henceforth, the energy that is light would be quantized.

With this one assumption, Planck was able to derive a mathematical formula that fit the data he had at hand quite well. To test his theory further, Planck invited Rubens to tea at his home outside Berlin. Rubens brought with him the results of his most recent measurements. Planck reciprocated with his new formula for the distribution of black-body radiation. For the first time, theory agreed with experiment completely. The success had all stemmed from Planck's bizarre assumption about the discrete motion of a pendulum.

Planck must have intuited the profound significance of this moment, because in the flush of this success he took his beloved son Erwin on a long and memorable walk in the Grünewald forest outside Berlin. Erwin, only seven at the time, knew of his father's long preoccupation with the analysis of light. As they walked the father confided to his child that he felt the discovery he had made was destined to be of the same significance as those of Copernicus or Newton. It was the stuff of revolutions. These were bold, prophetic words from an upright, cautious man, a heresy uttered to no one other than his son, and yet they were spoken truly. The paradoxical demands that Planck's theory of light has placed on the human imagination continue to reverberate a century later in the form of wave-particle duality.

Planck himself was reluctant to accept the quantum of light he had invented, struggling time and again to avoid the need for it. One of the very few to take up the implications of Planck's analysis of candlelight was the still-unknown physicist Albert Einstein.

The Reckless Quantum

According to the assumption contemplated here, when a light ray spreads from a point, the energy is not distributed uniformly, but consists of a finite number of energy quanta that are localized in space, move without dividing, and can be absorbed or generated as a whole.[7]

Albert Einstein, 1905

Einstein's greatest accomplishments date to the remarkable period from 1902 to 1908 when he held the unpretentious position of patent-office technician. One extraordinary scientific paper after another was written by this unknown giant in the few hours between work and domestic duties. During this period, Einstein, always a bold and prescient thinker, daringly leaped beyond Planck to suggest in 1905 that light be considered a collection of independent particles of energy.[8] With this hypothesis, Einstein went on to explain other recent experiments that had also posed insurmountable difficulties for the classical wave theory of light. The immediate reaction to his suggestion, supporting calculations, and predictions was silence. Particles of light? The idea was appalling. It flew in the face of a century-long vindication of the wave theory. Huygens, Euler, Fresnel, Faraday, Maxwell—had not their efforts finally culminated in the enormously successful view of light as an electromagnetic wave? Not only did experiments on the interference of visible light support it, but now also others at much longer wavelengths, including the newly discovered infrared and radio waves. Undeterred, Einstein was, in his own words, "ceaselessly preoccupied with the incredibly important and difficult" question of the constitution of light. His research convinced him that "the next phase of the development of theoretical physics will bring us a theory of light that can be interpreted as a kind of fusion of the wave and emission [particle] theories."[9]

Albert Einstein receiving the Planck Medal from
Max Planck in 1929.

In these early years, one of America's greatest experimentalists, Robert A. Millikan, had been studying the emission of electrons from metal surfaces when illuminated by light. The details of this phenomenon formed a strong if not completely convincing argument in support of Einstein's particle view. Still, the particle view of light was profoundly objectionable to Millikan, so objectionable that even in the face of his own data, he called it a "bold, not to say reckless, hypothesis."[10] That such a hypothesis could explain his data was an embarrassment, especially as no one took it seriously. Even Planck himself, who after 1905 had come to know and respect Einstein for his other scientific contributions, was sharply critical of the notion of light being advanced by Einstein.

An example of the furor that attended the introduction of light quanta took place in 1909 on the occasion of an address by Einstein to the eighty-first meeting of German scientists and physicians in Salzburg. Planck, Rubens, Stark, and other physicists were in the audience. Einstein gave a masterful overview of the situation regarding light, the supposed ether, and the necessity for light quanta. On the basis of his penetrating analysis of key experiments, Einstein maintained that "our current foundations of the radiation theory must be abandoned." He went on to ask the question that was on everyone's mind. Would it not be possible to derive Planck's result, and explain the key experiments without the "horrendous-looking hypothesis of light quanta," as he called it? Could we not retain at least the classical beauty of Maxwell's electromagnetic fields for the propagation of light through free space, and treat only the material processes of emission and absorption differently? Einstein's response was a firm no. He went on to reason, starting from Planck's own result, that to escape by such a route was impossible. The detailed results from experiments on black-body radiation constrained one in the most tenacious way, forcing one to assume light quanta, he said.

The first to respond to the lecture was Planck himself. Planck was willing to admit that there was a quantum discontinuity, but he simply felt convinced that it could and must be located entirely in the atomic nature of matter, and not ascribed to light itself. How could particles of light give the enormous variety of interference effects that wave optics could explain so readily? How could a particle interfere with itself? It was a logical absurdity!

Einstein responded by suggesting that the light quanta need not interfere with themselves, but could interfere with other quanta as they traveled. A brilliant response; he did not know then that this way out was also blocked! If quanta existed, they were even more shocking in their behavior than he realized. In fact, Einstein was forcing the entire physics community into accepting a quantum theory of light, a theory he firmly rejected later in life. Photons, the light quanta first introduced by Einstein, *do* interfere with themselves, as we will see later on. The stakes were far higher than even the boldest thinker of the period understood. Even Einstein's view of light as energy quanta would prove inadequate. Planck's caution was well grounded. As he had prophesied to his son, it was a revolution in thought the likes of which had not been felt since Copernicus or Newton, and the implications were frightening.

—

THE BATTLE BETWEEN Einstein and Planck received reinforcements from another side. Parallel with the passionate investigation into the nature of light was an entirely similar, and equally significant, investigation into the nature of matter. One of the key figures in this research was the charismatic Danish physicist Niels Bohr. His penetrating study of matter drew him at first indirectly, and later directly, into the debate on light. The theoretical controversies in which Bohr participated will seem remote from our common experience of light, but in actuality they are not. In this instance, the connection is not by way of candlelight but through lightning and the aurora borealis.

Night Sky and Neon Signs

Lightning is the father of light.[11]
Franz von Baader

To the nineteenth-century nature philosopher Franz von Baader, light is born from a sheath of darkness, just as an infant departs from the mother's womb. In light there dwells God's realm, the creator's love; and in darkness resides the creator's anger. The force of God that liberates light from darkness is the force of fire, and the expansion of that light is the Holy Spirit. One and the same divine being eternally manifests both in regions above as beneficent light, and in regions below in the terrifying form of lightning.[12]

The release of light is always swift, whether from the spark of steel against flint, or a candle wick catching fire. Light is born suddenly, and in danger, out of darkness. To von Baader, lightning was, therefore, emblematic of the genesis of light; it was the father of light. Nothing stands in greater contrast to a flash of lightning as it rends the air than quiescent candlelight. If, as von Baader says, lightning is the father of light, then the unlit candle must be its mother. Yet lightning presents not only symbolic truths but scientific ones as well.

We can begin, as we did with the candle flame, by studying the spectrum of lightning. This is especially easy because no slit is needed before the prism; the flash itself already confines the light to a narrow if irregular route. The dispersed image of lightning looks very different from that of a candle flame. In place of a smoothly varying band of colors, sharp lines of color appear, one separated from the next in an apparently irregular fashion. These are the "spectral emission lines" of the elements that make up our atmosphere. As the powerful surge of electricity passes swiftly along the lightning's channel, each element—nitrogen, oxygen, etc.—is caused to emit light of a kind that is completely characteristic of the element. The universal black-body spectrum, which was oblivious to the material heated, is gone. In its place one sees a kind of light uniquely tied to the individual elements that make up our atmosphere. The spectrum is not that characteristic of a heated body, but that produced by the flow of electricity through gas. Phenomena such as these—and their laboratory counterparts—captured the attention of Bohr in Copenhagen. They were responsible for launching a second attack on the embattled electromagnetic conception of light.

Lightning is not the only natural phenomenon that shows a line spectrum. Another and much quieter phenomenon associated with the polar night also shows its spectral colors, and it also has a history rich in emblematic meanings.

—

WHEN VIEWED FROM several thousand miles away in space, the polar regions of the earth show an extraordinary oval of light many hundreds or even thousands of miles in diameter. In the north the oval is centered over the northwest tip of Greenland.

Over time the light ring grows and contracts, changing its color and shape as well as its size. To someone on the ground, the oval of light appears like veils or draperies of gentle, mobile, delicately radiant color that can last for several hours and stretch across the entire night sky. Growing, fading, and rippling, these lights possess a morphology entirely of their own and have evoked wonder and stimulated all manner of speculation over the millennia.[13] Since Greek times, they have been called the *aurora borealis*, which means "northern dawn." (There is also an *aurora australis* or southern dawn in the southern hemisphere.)

While the possibilities are infinite, a typical aurora might develop in something like the following way: The night is moonless, dark, cold, and clear. At first, one sees only a simple, broad white arc in the northern sky with a smooth lower edge, its ends never touching the earth. The lower border then develops folds and kinks, and colors now show themselves, turning the silent glow from white to yellow, green, or red. The once-tranquil arc begins to pulse and becomes irregular in shape. With luck, the observer, by now completely enchanted, might see the magnificent "corona," which appears like streams of rippling light spreading out from a common center high overhead.

In antiquity, Aristotle reported the appearance of such "chasms, trenches and blood red colors" in the night sky. Later, Plutarch repeated the report of an earlier fifth-century B.C. aurora that was visible in Greece for seventy-five nights. Sightings like these are rare in the Mediterranean, but in some far northern (or southern) regions an auroral display can be seen every clear dark night. Especially in these regions, the lore concerning the aurora is rich and magical, reminding us of the rainbow's power of enchantment.

Among the Eskimo, the aurora is associated with the souls of the dead. Those who suffered a voluntary or violent death cross a narrow and dangerous bridge and enter heaven through a hole. Once arrived, they play ball, and their game is seen below as the aurora. The Eskimo word for the aurora is *aksanirg*, meaning "ballplayer." The Eskimo of northern Canada called the aurora the "dance of the dead," while in Greenland stillborn or premature children are said to play ball in the heavens following their untimely death, and so give rise to the aurora.

In Northern Europe, the Finns tell the tale of Repu, a fox whose tail flashes fire but not heat—the aurora. The modern Finnish word for the aurora belies its origins in the tale of Repu; *revontuli* means "fox fire."

As a prognosticator, unusually dramatic auroral displays have been almost universally taken as omens of war and disaster. Even in the twentieth century, the interpretation of the Miracle of Our Lady of Fatima, when the Virgin appeared to three children on May 13, 1917, is connected with the aurora. The Virgin

The aurora borealis.

told the children "when you will see a night illumined by an unknown light, know that the chastisement of the world is at hand." On January 25, 1938, an unusually brilliant auroral display appeared that was widely taken to foretell Hitler's invasion of Austria three months later.

—

WHEN THE FINNS held that Repu's tail flashed bright fire but not heat, they were not far off the mark. Unlike the light from a candle flame or other incandescent sources, the light of the aurora is not due to the glow of a heated body, but is a "cold" light much more akin to the light given off by a neon sign or Galileo's solar sponge, than by a candle. Like lightning, auroral light can be sent through a slit and prism to produce spectra. It, too, shows discrete luminous lines similar to those seen in lightning spectra, and both are like those investigated by Kirchhoff and Bunsen.

The physics behind the brilliant red light of a neon sign can help us understand the light of lightning and aurora. In both cases their light is produced by electricity passing through a rarified gas atmosphere. If the gas is neon, the light is red. Other gases give other colors when electricity passes through them, and as Kirchhoff and Bunsen verified, the spectrum of each is unique. The source of electricity for a neon sign is obvious, but if the aurora and lightning are similar, where are the dynamos that power them?

If, as he watched for sunspots, Galileo had simultaneously studied the night sky in Norway, he would have noticed a correlation. The intensity of the auroral display corresponds to sunspot activity with a two-day delay between them. The blemishes of the sun are huge eruptions that send an energetic stream of charged particles racing toward the earth.

That stream is channeled by the magnetic field of the earth to the poles, and the aurora shines out when that dark solar

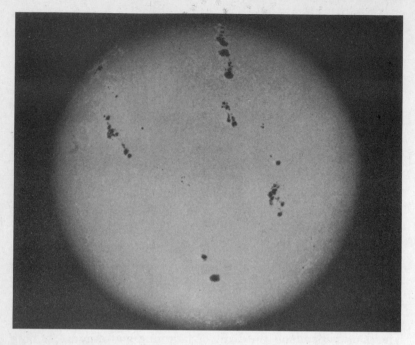

Sunspots, the unsightly imperfections discovered by Galileo.

stream strikes the earth's upper atmosphere. The earth's invisible magnetic forces draw down a dark solar wind, brightening the long polar night with fox fire.

Planck's magisterial analysis of light alone, even with the hypothesis of the quantization of light energy, is insufficient to account for lightning, or auroral light. Something more is needed, namely a new theory of matter. The aurora poses new questions concerning the structure of the atom and its production of light. For this we turn to Niels Bohr.

—

BORN IN DENMARK in 1885, Bohr left Copenhagen for England in 1911, immediately after receiving his doctorate. He first joined Sir J. J. Thomson (discoverer of the electron) at the Cavendish Laboratory in Cambridge, England, and then went on to work with Sir Ernest Rutherford in his laboratory at Manchester. Rutherford was busy at the time with his famous alpha-particle experiments that would lead him to a picture of the "nuclear" atom in which most of the atom's mass is concentrated at its center, in a positively charged nucleus. Fresh from collaboration with these men, Bohr returned home in 1913, and shortly afterward turned his attention away from nuclear structure and toward the problem of light.

Study of the spectral lines of the elements had, by 1913, developed into an active area called spectroscopy. Thousands of lines crowded spectroscopic tables without any apparent order. One day in casual conversation with a colleague, Bohr learned that the physicist Rydberg had found a formula that predicted many of the prominent spectral lines, but it completely lacked any theoretical base. He had discovered a pattern within the jumble, but still no reasons existed for it. Bohr quickly recognized that by combining Rutherford's nuclear atom with the Planck hypothesis of discontinuity, he could derive Rydberg's formula and so explain the detailed appearance of line spectra, such as appear in the aurora or lightning. But as with Planck, the derivation came only at a very considerable cost to the presumptions of nineteenth-century physics. Bohr's atom was to be understood as like a planetary system with electrons circling a central nucleus in concentric orbits. This much could be thought of along traditional lines. The transitions, however, from one orbit to another needed to take place discontinuously as "quantum jumps." With each quantum jump, a quantum of light or photon was given off. Its color or frequency could not be given by a classical analysis of the atom's motion in any sense, but by Planck's famous relationship: $E = h\nu$.

Physicists in England and Germany responded to Bohr's suggestions incredulously, but as James Jeans remarked to colleagues, although it might have nothing else in its favor, it did possess the "very weighty justification of success." Planck's quantum was irrepressible. Here was an entirely new domain of application for the idea. With Bohr's theory much could be explained, even if no detailed mathematical theory yet existed. When the young Werner Heisenberg from Munich later succeeded in developing such a theory, Bohr's planetlike, electron orbits disappeared from the atom, which became a stranger object still. Very little of the traditional, mechanistic world conception seemed left standing. Atoms were, after all, the fundamental building blocks of all matter, and yet nearly every recognizable feature of them disappeared before careful scrutiny. The electron has no orbit, no real position in the atom; light is emitted during discontinuous quantum jumps; how was one to think such a world? It seemed impossible!

When Einstein first heard Heisenberg present his new theory of quantum mechanics in the famous Berlin colloquium series hosted by Planck, he was shaken.[14] Afterward, the great man invited Heisenberg home to talk. Immediately, Einstein complained that in Heisenberg's formulation the concept of the "electron path" had no place! Heisenberg, who had read Einstein's work on relativity carefully, quoted Einstein back to him, pointing out that since we cannot, in fact, ever observe such a path, it made no sense to introduce the concept into the theory. (Einstein had made similar arguments when advancing parts of his relativity theory.) Much to Heisenberg's amazement, Einstein replied, "Perhaps I did use such a philosophy earlier, and also wrote it, but it is nonsense all the same." If the loss of a path for the electron was bad, the notion of a quantum jump was, if anything, even less satisfactory to Einstein. There seemed to him no physical basis for understanding Heisenberg's quantum theory. It provided a formalism and predictive power, but no explanatory value in the usual sense. Against this he rebelled for the remainder of his life. The boldest thinker of the decade balked once the full implications of quantum theory, which he himself had helped create, were known. Bohr, by contrast, began cautiously but gradually to assume a more and more radical position as experiments suggested it.

—

BOHR WAS ENTIRELY Einstein's peer, and a man of enormous charm and personal force. When Einstein finally met him in 1920, he felt the power of Bohr's personality: "Not often in my life has a human being caused me such joy by his sheer presence . . . [and to the physicist Ehrenfest] I am as much in love with him as you are. He is like an extremely sensitive child who moves around the world in a sort of trance."[15] Bohr, like Planck, was reluctant to give up the great beauty of classical field theory for an atomistic conception of light. For twenty years, Bohr resisted the concept of particulate light with every weapon in his arsenal, giving up even the sacred notions of causality, and the conservation of energy and momentum in individual events, if only he could avoid the conception of light quanta. He, again like Planck, argued that the locus of the problem was not light, but matter. The magnificent edifice of electromagnetic theory should not be modified, but rather our conception of substance. It should be reconceived along lines advocated in his famous paper of 1913.

What one detects as a granularity of light is, he argued, not a feature of light but only the signature of its origins. An analogy can make this clear. A wonderful yet ridiculous photograph of Einstein exists, taken during a visit he made to the American southwest. It shows him in a broad-brimmed sombrero and poncho, standing in front of an adobe hut surrounded by native southwest Indians. He is grinning impishly, his unkempt white hair sprouting out from under his hat, and his eyes twinkling. He was in the southwest, but no matter how you dressed him, or in what company you put him, he was still Einstein. Similarly, when light is emitted from matter, it carries away with it the costume of the source. Spectral lines, for example, clearly reveal, in the arrangement of their colors, the particular element from which they were emitted. One should not, however, confuse the present dress for the essence of free light. The being of light may bear the imprint or signature of the source, but light left to itself is a continuous wave, the argument went.

Bohr's view, one which has persisted with many successes until this day,[16] conceived of matter as atomic and quantized, but insisted that light as it traveled between emitter and detector, that is when left to itself, was a pure electromagnetic wave.

The drama heightened when truly convincing experimental evidence for light quanta appeared in 1922. A. H. Compton scattered very high energy light in the form of X rays from free electrons. The effect was immediately intelligible if one thought of the collision between the light and the electron as if it were a collision between two billiard balls, one being the electron and the other a quantum of light. The wave theory seemed hard-pressed to give any account of the effect at all. The tide was clearly turning, and an atomistic conception of light was gaining ground. As the superb Munich physicist Arnold Sommerfeld put it, Compton's discovery "sounded the death knell of the wave theory of radiation."

Still, Bohr resisted. Together with Kramers and the young Harvard graduate Slater, a theory of the Compton effect was devised that saved the view of light as a wave. However, in order to retain the wave theory, Bohr made an enormous concession. He gave up strict causality, as well as the conservation of energy and momentum for individual quantum events. Conservation would occur only on average, statistically, and causality only overall. This was an extraordinary compromise consciously made, one Einstein protested adamantly when he heard of it. Causality had been sacred for centuries, how could it be thrown out in any way whatsoever? So, too, with the conservation of energy.

The theory of Bohr, Kramers, and Slater could be tested by a refinement of Compton's experiment in which X rays are scattered off electrons—it failed. Although his theory failed, Bohr had dared to challenge the bedrock classical conceptions of causality and energy conservation. Once sounded, the challenge refused to die. Was the world a causal chain from beginning to end, or did quantum theory require a looser conception of necessity? In the first decades of the century, these were questions debated in the widest circles, far beyond the confines of physics laboratories.

Living Philosophy

Berlin at the turn of the century was a heady place. While Max Planck was occupied at the Berlin University with the invention of quantum theory, not far away at the old Café des Westens, prominent writers, artists, and intellectuals gathered every evening to discuss Van Gogh, Nietzsche, Freud, and to sow the seeds of revolution. The poet Ernst Blass experienced life in these years as "a spirited battle against the soullessness, the deadness, laziness, and meanness of the philistine world."[17] Nineteenth-century materialism was under attack on every front. Quantum theory and relativity would undermine mechanistic explanations, artists were discontent with the constraints of academic art, and intellectuals were announcing a countercultural revolution in ideas.

New attitudes rejected the arid mechanisms of the previous age, and longed to replace them with a more vital *Lebensphilosophie*, or living philosophy of the world. Physics, never performed in cultural isolation, shared in the spirit of the times. One eminent professor of physics said in a 1925 address, "It is interesting to observe that even physics, a discipline rigorously bound to the results of experiment, is led into paths which run perfectly parallel to the paths of the intellectual movements in other areas of modern life."[18]

Together with the revolutions in physics occurred a revolution in sensibilities, but the winds of change did not affect everyone equally. The first quarter of the twentieth century was often a pitched battle between the voices of tradition and those of a new life. Our chronicle of light reflects these struggles. Planck, Einstein, and Bohr will fight to protect the traditional underpinnings of science while simultaneously advancing revolutionary understandings of light. Artists and spiritual thinkers of the period, while sometimes appreciative of science's grand accomplishments, nonetheless criticized its one-sidedness and longed to incarnate Emerson's dream, that is, to unite severe science with a poetic and spiritual vision.

In Paris, Robert Delaunay painted directly out of color and light, juxtaposing color forms with only a hint of realism. In 1912 Klee translated Delaunay's essay "On Light" for the expressionist art magazine *Der Sturm*. Delaunay's words in that essay could have been Klee's or Kandinsky's: "So long as art is subservient to objects, it remains description, literature. . . ." Rather should light itself be "treated as an independent means of representation."[19] These artists turned their trade away from depiction of sense objects and toward what had, until then, remained unrepresented and extrasensory—the lights of nature and mind.

Thus, with the dawn of the twentieth century, artistic and spiritual currents were astir that reanimated and recast the ancient and sacred lineage of light. Artists courageously explored the new light's first feeble glimmerings, while spiritual philosophers such as Steiner pursued light along pathways of meditation. The range of views alive during the first twenty-five years of the century was incredible, and the relationships between them were sometimes explosive. We have inherited all it attempted, both cultural and countercultural.

—

LIKE A RHIZOME whose tendrils stretch out from invisible immortal roots, the metaphysics of light had lain buried beneath the nineteenth century's scientific achievements; yet it had gone on working still. When the soil overhead grew thin, an unsuspected stem reached upward, toward the light, and blossomed at the dawn of our century in the arts, literature, and in visionary philosophy. W. B. Yeats's words could well have been written to describe the renewed inspiration offered by a metaphysics of light among artists, poets, and philosophers of the time.

> The shapes of beauty haunting our moments of inspiration [are] a people older than the world, citizens of eternity, appearing and reappearing in the minds of artists and or poets . . . and because beings, none the less symbols, blossoms, as it were, growing from invisible immortal roots, hands, as it were, pointing the way into some divine labyrinth.

Until the First World War, the two worlds of light—science and spirituality—develop in relative isolation from each other. Following the war's devastation, however, they confront one another with renewed intensity. Some will plead for union, others for segregation, and the figure of light will bear the impress of those controversies.

Crippled by the devastation of the First World War, and the harsh terms of the Versailles treaty that ended it, Germany's economy was in a state of collapse. Many of the most talented men of the new movement in science, art, and religion had perished in the trenches. Among the ruins of Europe, living philosophy gained momentum. The flames of discontent were, however, also being fanned by early factions of Hitler's fascist corps, who fed on the popular feelings of resentment. Reactionary sentiments grew and Hitler managed to divert some of the countercultural movement's impetus away from its original intentions and used it for his ends. Prominent individuals unwittingly added their own reasoned voices to the growing tide of repression.

The smoke of many fires darkened the sky. During the years of Hitler's rise to power, Einstein's theory of relativity was labeled a "Jewish theory" and Einstein resigned his position at Planck's Kaiser Wilhelm Institute in Berlin. Planck, von Laue, and Heisenberg maintained a precarious existence within the German scientific community during the Third Reich, struggling to retain a shred of humanity in an increasingly insane world. During the war years, Planck, now very old, took up lecturing tirelessly on his understanding of the relationship and division between science and religion. Still, the furnaces of hatred belched forth their poisonous vapors. Planck watched helplessly as his own Kaiser Wilhelm Institute was "cleansed" of Jewish scientists, and then used to advance Hitler's dream of a pure Aryan world order. The abstract ideal of science disconnected from the values of its society fell apart. Good science is good not because it is value free, but because it cherishes noble values.

On February 15, 1944, a massive Allied bombing raid on Berlin obliterated the outlying town of Grünewald and with it Planck's home. All his meticulously preserved letters and documents were destroyed in the conflagration. Finally, later that same year, his beloved son Erwin, with whom he had walked alone in the Grünewald forest at his first expansive moment of discovery, was arrested by the Gestapo for his role in the assassination attempt on Adolf Hitler. His father moved heaven and earth to save Erwin's life, to no avail. The pain of his loss nearly killed him. It was a very, very dark time. Perhaps, as von Baader felt, out of such darkness, coming swiftly like a bolt of lightning, light can rise again.

CHAPTER 10

Of Relativity and the Beautiful

It seems that the human mind has first to construct forms independently before we can find them in things.

Albert Einstein

Unlike Planck or Bohr, Albert Einstein began his study of light not with a candle flame or the aurora, nor with any outer phenomenon, but with a *Gedanken* experiment—an experiment he could perform purely in thought. Einstein would conceive many such thought experiments, trusting in each to his inner sense for the truth they contained. In contrast to many skeptics, he dared to believe "that pure thought can grasp reality, as the ancients dreamed."[1] In his first thought experiment, Einstein locked horns with light. So elusive to laboratory experiments, perhaps pure thought would have better luck in grasping the reality of light.

Einstein himself tells us that probably in early 1896, as a sixteen-year-old, he had the impossible thought; if one runs after a light wave with a velocity equal to that of light, then the light wave should stop. Traveling with a friend, you do not notice the speed of the plane carrying you both. By running with light, likewise, it would seem to stop. "However," noted Einstein, "something like that does not seem to exist!"[2] He had arrived at a marvelous paradox, one he could think, but to which he saw no solution. One can imagine running at any speed, and so accompanying any moving object. Yet light was different; its travel seemed somehow special. Unmoving light? Such a thing was impossible! Yet by thinking the impossible, a revolution was born.

The paradox of running with light would simply not let go of Einstein's imagination. Throughout his years as an undergraduate student, he turned it over and over in his mind, longing to unravel the mystery of light, and together with it, the nature of the supposed ether that carried light. For nearly ten years he harbored the puzzle, vainly suggesting experiments to his instructors, grappling with his paradox. During those years he matured scientifically, and like the grain of sand in the growing oyster, his question concerning the nature of light was the irritating nucleus around which grew the pearl of relativity. Finally, in 1905, during a year of amazing productivity, the being of light finally revealed itself to him, at least partly. Under his steady scrutiny a new thought form was shaped, one capable of addressing his early question; it has become known as Einstein's special theory of relativity. The paradox of running with light found an answer. Together with that answer came innumerable predictions, many of which startled Einstein's audience, and continue to startle us still.

Planck had been reluctant to suggest that light might be quantized into what later would be called photons. Unlike his Prussian counterpart, Albert Einstein's thinking was more daring. He felt unconstrained by either traditional family values or the habits of thinking characteristic of classical physics. When the celebrated French mathematician Henri Poincaré wrote a letter of recommendation for Einstein in 1911, he emphasized his uncanny openness to new ideas, and his extraordinary ability to see all they imply. "One can, above all, admire in him the facility with which he adapts himself to new concepts and draws all the consequences from them . . . he is not one attached to the principles of classical physics."[3] Indeed, as we have seen, Einstein was one of the first to take Planck's work seriously and quickly to develop its implications.

Others at the close of the nineteenth century had been groping toward a new view of light, space, and time, but in every case, some vestige of the classical conception held them back. They could not free themselves from a mechanical image of light, nor from the framework of absolute space and absolute time. Faraday's ontology of force had been "solidified," by accommodating it to ether theories. Throughout the nineteenth century, science viewed light as a vibration in the universal, material ether. It was a view that was failing, but still proved amazingly persistent. Early on, Einstein recognized the problem and wrote, "the incorporation of wave-optics into the mechanical picture of the world was bound to arouse serious misgivings."[4] One such misgiving must certainly have arisen from Lorentz's proof at the end of the nineteenth century that the rigidity of the ether would not merely need to be large, but *infinite,* in order to support the vibrations of light. As Einstein remarked, "The laws [of Maxwell] were clear and simple, their mechanical interpretations clumsy and contradictory."[5]

As a consequence of such considerations, physicists on the continent began to dematerialize the ether, stripping it of unnecessary physical attributes. Important steps were taken by the period's recognized master, Lorentz. He eliminated every mechanical feature of the ether, save one. Gone were its mass and resilience, but it still defined an absolute frame of reference. It was immobile. All motion was to be measured relative to it. From this frame, one had a "God's eye view" in which all things could be seen as they really were. Einstein, but twenty-six years old, questioned even this. He worked inwardly, theoretically, forging new ways of thinking about old things such as light, the ether, space, time, and causality. Once the interior forms were perfected, they could be compared to outer phenomena to see if they could withstand the stringent demands of experiment.

Of the many rich veins we might follow into Einstein's theory of relativity, none glitters more brightly than light. As we follow it, two revolutionary features of the theory will quite naturally occupy us: the complete collapse of the hypothetical ether, and the unique significance of the speed of light. They reflect the two postulates, as Einstein called them, on which the special theory of relativity was founded.

Relativity and Faraday's Archetype

For the rest of my life I will reflect on what light is!
Albert Einstein, c. 1917

In the opening paragraphs to his famous 1905 paper, Albert Einstein moved dramatically beyond the views of British physicists, and went further even than his continental colleagues, when he wrote, "The introduction of a 'luminiferous ether' will prove to be superfluous. . . ."[6] Not only did Einstein suggest that previous scientists were wrong about their substantial ether, but even the totally ephemeral ether of the Dutch physicist Lorentz was unnecessary. In that remarkable paper, Einstein successfully developed a framework for physics that did not possess a unique frame of reference against which to measure motion. *All* motion was to be judged only with reference to other objects; no viewpoint was privileged with the designation of absolute rest. Moreover, *within each frame of reference, all of the laws of physics applied equally well*. This was Einstein's first postulate, which he called "the principle of relativity." To someone at home in one frame of reference, everything was explicable according to the very same laws used in every other inertial frame. It seemed an innocent assumption, yet it led to radical changes in our understanding of the world, including the ether. Since all frames were perfectly equivalent, there was no need for a single absolute frame of reference provided by the ether.

In order to introduce his pivotal principle of relativity, Einstein turned to the archetypal phenomenon of electromagnetic induction discovered by that fertile experimenter Michael Faraday. This same experiment had first implied a possible connection between light and electric disturbances for Faraday, already a revolutionary association. Now it was pressed into service by Einstein for a second and equally significant end. Recall that when a magnet is moved relative to a coil of wire, a current arises in the coil. But, asked Einstein, what is *"really"* moving in such an experiment, the magnet or the coil?

To make the situation clearer, let us perform a thought experiment. Imagine a doughnut-shaped space station floating freely in space. Its metal skin is a good conductor of electricity. Toward it speeds a spaceship, ordinary in all things except that it has been made into a huge, cylindrical magnet. One end of it is a north magnetic pole, the other a south pole. The magnetic spaceship approaches the station and flashes through the central hole of the space station at great speed. On the station a large current surge is detected in the metal skin. Why? According to Faraday's law, the surge is due to an *electric field* induced around the circular station by the moving magnet. Calculations are made by space-station scientists, and yes, experiment agrees with theory. But now shift to the viewpoint of those aboard the spaceship. They have not been moving at all, and toward them rushes the space station. As it flashes by they also correctly predict a current surge in the station of exactly the right kind, but their explanation of its origins differs completely from that of scientists on the space station. To those in the ship, the current is due to the movement of charges (electrons) in the skin of the station acted on not by an electric field (as those on the station maintain) but purely due to *magnetic forces*. Who is correct? Did the current arise by the action of electric or magnetic forces?

Clearly, there is no way to tell. Nor does it matter for practical purposes which view we take, because in either case a current arises. There is, however, an enormous difference in what we say causes the current. From one perspective, Maxwell's electromagnetic theory credits the current to the action of a *magnetic field*, while in the other case, it arises from the action of an *electrical field*. That is, a physicist sitting in the space station (which plays the role of the coil in Faraday's induction experiment of chapter 6) will give one explanation for the current (calling it an induced electric field effect), while a second riding on the magnetic spaceship will give a very different account (saying it is due to the motion of charges in a magnetic field). They will both be using the same theory of electromagnetism, but the explanations they offer will be fundamentally different. How can this be?

Faraday's coil/magnet experiment provides us with a beautiful, indeed, an archetypal instance of relativity. What one observer judges to be caused by electric fields, another observer in motion relative to the first will judge as due to magnetic fields, or some combination of both. Faraday's simple experiment is, in the hands of Einstein, the basis for another revolution in our understanding of light and causality. First, he will use it to dismiss absolute motion and with it, the ether. Second, it requires us also to view electric and magnetic fields as related relativistic quantities. They are not "out there" like some kind of surrogate ether, but transform into one another according to Einstein's equations. Accounts will, therefore, differ for physical effects such as current production, but a good theory (like Maxwell's) can be used in any reference frame with equal success.

Just this universal applicability makes the ether unnecessary. The ether had always provided the absolute frame of reference against which all motion was to be judged and the "true" theory applied. Now one no longer needed to say what was "really" moving, the magnet or the coil; it just doesn't matter, so long as one is willing to give up the goal of a *single* true account for physical events. Each observer can give his own perfectly consistent causal account of why things happen. However, what each says "causes" an event (for example, the current in the coil), or even the timing of the events, these will now depend on their relative states of motion.[7] There is no "God's eye view." Accept this, and the accession is a major one, and all preferred frames vanish, including that offered by the ether. That persistent artifact of the materialistic imagination, the ether, is finally gone!

Wait! Einstein might have no need for the ether to explain electrical and magnetic effects, but what then is light? All wave theories supposed it. Maxwell and those following him took light to be an electromagnetic vibration in the luminiferous ether, propagating according to his four equations. If light is a wave, but there is no ether, then what is waving? Try to imagine sound without the motions of air. Yet this is just the avenue Einstein was offering for light. Einstein's suggestion received experimental support during these same years. The consequences for our scientific understanding of light are enormous. Light is to be an electromagnetic wave, but without a material medium to support its motion. Light's last vestige of materiality, its physical body so to speak, had been taken from it.

With the ether gone, what remains? Perhaps the electric and magnetic fields themselves are real even if the ether is not. No, Faraday's archetypal experiment and the principle of relativity clearly demonstrate that it is wrong to think of the electric and magnetic field at a particular time and place as having a single true value. Observers in motion with respect to one another measure very different values for the fields at the same point.[8] Faraday's field concept gets us closer to light, but in using his idea we cannot separate it from ourselves. Our account will reflect our bias that we are stationary and the world moves around us. Ptolemy took his viewpoint on the universe to be the earth, Copernicus chose the sun; the accounts they gave of the motions of the heavenly bodies were equally accurate but *very* different. Likewise, relativistically we are always implicated in every measurement. Physicists must always specify what frame of reference they are assuming for their calculations. It does not matter in principle what frame they choose (although in practice one frame is usually much easier to work in than another), but the story they tell will be colored throughout by the choice. If one forgets this lesson from relativity, then one runs the danger of slipping back into a naïve realism that takes one view and assumes it to be true universally.

Alternatively, we might turn hopefully to quantum theory for consolation. If light waves are now problematic, then return to a corpuscular view. But while quantum theory speaks of a particlelike nature to light, the photon remains massless and possesses, in addition, its own set of unusual properties. It implicates the observer even more deeply through Heisenberg's uncertainty relations. No, quantum theory can deepen the mystery, but it cannot relieve it. Light is not substantial, at least not in the form of a massive particle à la Newton, nor even in the form of undulations through a material medium à la Euler or Maxwell. If one conceives of the universe as matter or its movement, light is the exception that shatters that prejudice. The nature of light cannot be reduced to matter or its motions; it is its own thing.

Pause to consider the history of light once again, now including the work of Einstein, for we have come far. First experienced as the gaze of God, light was a spirit reality, eternal and omnipresent. From the time of the Greeks to those of Newton, light became first geometrical and then material. With Faraday and Maxwell the era of electromagnetic light dawned, but light still carried along with it the trappings of its recent past, a material medium—the ether. Aside from Faraday, scientists simply could not imagine light—a real thing—apart from materiality. Their thinking did not allow it; matter was synonymous with reality. Then in 1905, Einstein suggested an alternative. The cost was high: it granted no absolute meaning to motion, declared the ether superfluous, allowed no single privileged causal account for physical phenomena, implicated us in all observations, and predicted a host of attendant effects such as length contraction and time dilation. Yet for all that was lost, Einstein provided much in return. If there were now many causal accounts, he gave physicists the means for translating one to another, accomplished by what is called a Lorentz transformation. Later, he also offered the basis for seeing a deep connection between matter and energy $(E = mc^2)$. Although not material, light could be thought of as a form of energy whose behavior in this world is imaged in Maxwell's electromagnetic field equations. Rationality remained, simple or subtle matter did not.

—

COPERNICUS HAD FORCED human consciousness from its ancient haunts to the sun. The earth was no longer the still point around which the cosmos moved; the earth itself moved around the sun, and it was but an insignificant star in the galaxy, which in turn was but one among countless others. No single material place was the center and focal point of God's creation, but all were created equal. Einstein went further. Now there was not even a place of rest, everything everywhere was equally in motion. With Copernicus, location had become relative; with Einstein, motion. Together they emancipated human consciousness, although many longed for the security of the earlier view. One by one the traditional, outer props of science, religion, and society were being taken away. Humanity now stands alone and alienated in the limitless universe. Bereft of parents and far from home, each of us has now to become a center unto himself or herself, and find the spiritual strength to hang unsupported in the void, drawing support from within not from without.

Amid this extraordinary flux of new ideas, the loss of absolute space, time, and the ether, there remained but one constant of unique significance, one "truth" that remained independent of all reference frames—the speed of light! Around it swirled all other velocities, positions, and times, but light was, as Einstein wrote in 1905, "always propagated in empty space with a definite velocity c which is independent of the state of motion of the emitting body." This was Einstein's second postulate and pillar for his theory of relativity.

The Speed of Sight

No corporeal substance can be so subtle and swift as this.
William of Conches,
12th century

The night sky of ancient Greece did not suffer from the obscuring elements of city lights and pollution. In town and field, the stars were a nightly presence whose beauty and order were the springboard of science. The earliest measurements of the speed of light were made by observers who, like so many before and after them, turned their eyes to the stars and wondered.

If, as Empedocles, Plato, and many others held, during the act of sight something passes from the eye to the object seen, then in seeing the distant stars, something must pass through the greatest imaginable distance, and do so in an instant. After all, on opening our eyes, the entire world from horizon to horizon, and out to the most remote star or planet, immediately springs into view. Whatever travels there must do so at the highest speed. The measurement implicit in this argument, a common one in antiquity, was not actually a measurement of the speed of light, but really of the "speed of sight." The concern of ancient authors with the speed of perception continued for two thousand years. John Pecham, the most widely read medieval optician, wrote only of it, and not the speed of light, in his important work *Perspectiva Communis*. Still the speed of light itself was treated implicitly or explicitly from Plato on.[9]

Aristotle berated Empedocles and Plato for their views of sight.[10] In his view nothing streamed from either eye or object. Light was, as we have seen, a state or quality of the medium, and so just as water may freeze all at once, the transformation of the "potentially transparent" to the state of actual transparency, which is light, can occur everywhere simultaneously. It is simply wrong, according to Aristotle, to think of light as propagating at all. If his position is difficult to understand, consider a student who is, as are many, confused by Aristotle's claim that light does not propagate. Call the student's state of confusion darkness. As the instructor, firelike in his enthusiastic explanation of the puzzle, goes on, the student's face suddenly "lights up." The student has changed from a state of confusion to one of comprehension, and has done so instantly. Similarly, when the world "lights up," propagation is not a logical necessity. Light may well be a universal change of state.

The general view that light travels at infinite speed was taken up, for example, and repeated by Augustine: a ray of light "manifestly passes through those great and immense spaces immediately, in a pulse beat," he wrote. Or elsewhere, "our visual ray does not reach near objects sooner and distant objects later, but passes through both intervals with the same swiftness."[11] Bear in mind Augustine's close association of God and Christ with light. It made eminent theological sense that God's presence, near and far, be simultaneous.

By the twelfth century, Plato's influence had made its way to the Chartres masters, influencing their understanding of light and vision. According to William of Conches, the visual fire flowing from the eyes is subtle and swift: "no corporeal substance can be so subtle and swift as this. Thus it must be instantly here and instantly there [among the stars]." Still, William reasoned that because the visual fire was, in his view, a material substance, its movement required time. What is divine can travel at infinite speed, and so be all places at once, but what is material must travel through space and in time. Once again spiritual and physical considerations intertwine.

While light was taken by most to travel instantly throughout space, dissidents like William of Conches felt that although light might travel swiftly, it still required time to do so. Alhazen and Roger Bacon were two of the earliest to hold this opinion. By the time of René Descartes, the controversy was well established, with Descartes taking the side of instantaneous propagation in the strongest possible terms.

Recall Descartes's *plenum* and his model of vision in which sight is explained by analogy to a blind man with his cane. If all space is filled with a material medium, then, as with the blind man's rigid stick, a shock or movement at one end will instantaneously appear as a shock or movement at the other end. Descartes felt so strongly about this point that he was willing to hang the truth or falsity of his entire philosophical system on the prediction. To an unknown friend he wrote in 1634 that light "reaches our eyes from the luminous object in an instant; and I would even add that for me this is so certain, that if it could be proved false, I should be ready to confess that I know absolutely nothing in philosophy."[12] In Paris, forty-one years later the Danish astronomer Ole Rømer demonstrated that Descartes was wrong. The completion of the syllogism is left to the reader.

By careful observation of the moons of Jupiter, and using Cassini's new measurement of the planetary distances from the sun, Rømer was able to show that light travels at a finite, if extraordinarily high speed. We have already seen the significance of that speed for Maxwell, leading him to suggest that light was an electromagnetic wave. Its significance for the special theory of relativity was greater still. In the words of the French physicist Marie-Antoinette Tonnelat: "The second postulate of Einstein profoundly modifies the status of the speed of light: what was a kinematic and essentially relative entity now becomes a phenomenon describable by an invariant law."[13]

The goal of ever more precise measurement of this remarkable quantity, the speed of light, has captured the imagination of great experimentalists for the last three hundred years. The American physicist Michelson dedicated the last twenty years of his life to ever more exact measurements of light's speed of travel. His efforts culminated in a huge experiment involving a three-foot-diameter steel tube one mile long. Sealed and evacuated, light flashed back and forth through its airless corridor. The results—a new value for the speed of light—reached the impatient Michelson on his deathbed on May 7, 1931. He could die, two days later, content.

Since then ever more precise methods have been devised to measure the speed of light. The last used highly stabilized lasers and special techniques for measuring ultrahigh optical frequencies. Measurement finally became so exact that the main inaccuracy in them was due to uncertainties in the world's standard unit of length: the meter. In 1983 it was therefore decided to end the three-hundred-and-eight-year history of the measurement of the speed of light forever. Instead of defining a unit of length, as had always been the case before, the speed of light would be defined! Its value was taken to be the then-current best-measured value of 299,792,458 meters per second. From 1983 on, the speed of light was no longer a quantity that could be measured; rather it was defined as a matter of pure convention to be the above value.

Having given up the definition of the meter for that of the

speed of light, physicists had to go back and adjust the unit of length so it would be in agreement with the new standard. The meter, therefore, is now defined as the distance traveled by light in vacuum during a time interval of 1/299 792 458 of a second. The carpenter's measuring tape is really so many light-seconds long. Every ruler can now trace its pedigree back to the speed of light. As Grosseteste imagined, light, as it spreads, gives rise to space.

—

THE ANCIENT QUESTION of the speed of light has been answered. It is finite, and we have decided to call that finitude 299,792,458 meters per second. Never again will a researcher measure its value. What had begun as the infinite speed of sight has become the finite speed of light. Yet while experimentalists resolved one issue, relativity theory opened another, taking the great but finite speed of light, and declared it unique.

The Uniquely Universal Speed

We shall find in what follows that the velocity of light in our theory plays the role, physically, of an infinitely great velocity.[14]

Einstein, 1905

Drop a small stone into still water and watch the ripples expand. When the waters have quieted, skip a rock on the water's surface, and each time the stone touches, an expanding ring of waves grows outward. The speed at which the rings expand over the water's surface is exactly the same in both cases. The speed of water waves does not depend on the speed of the stone, dropped or tossed, but is determined only by the properties of water. Similarly with sound; the air alone, not the source speed, determines the speed of sound. All waves propagate at speeds set solely by the media that support them. Therefore, when Einstein put down his second postulate—"the speed of light is independent of the speed of its source"—it might seem that he was saying nothing new. In 1905 (almost) everyone thought light was a wave, so to them his second postulate was obviously true. But as we have seen, in that same paper Einstein had dismissed the ether; the very basis for the unique determination of the speed of light was therefore gone! Until 1905 wave velocities could be independent of source velocity *only* because they were completely fixed by the media in which they moved (water, air, ether, etc.). With *no* medium left for light waves, how could light possibly "know" how fast to travel? Einstein's reply was startling. He entirely ignored the question, and asserted as a postulate that the velocity of light was independent of source velocity. Period. When this is taken together with his first postulate, the principle of relativity, the consequences of his theory for our conception of time and space were truly revolutionary.[15]

Think back to Einstein's first thought experiment—running with light. What the second and first postulates imply is that not only can you never catch up to a light wave, but you can't even get close. Give your friend a flashlight, turn it on, and measure the speed of light: 299,792,458 meters per second. Ask him to keep measuring, just to make sure the speed of light does not change. Now start running. For argument's sake, let us assume that relative to your friend with the flashlight, you are running at ninety-nine percent the speed of light. Not knowing relativity, he will think you have almost caught up to the light wave. While running, you reach out and measure the speed of light yourself: 299,792,458 meters per second! How can this be? You are running at nearly the speed of light, and you measure the light as traveling at the same speed as before. You reason, it must "really" be going at twice the speed of light. Your friend with the flashlight, however, keeps getting the same value you do. What is going on?

Nothing else travels the way light does. All observers will

measure its speed in vacuum to be 299,792,458 meters per second. To make this possible we must change our conception of space and time, but if we do so, everything hangs beautifully, if strangely together. Is light really like this? Perhaps Einstein's theory is brilliant but wrong; theories have been disproved before. In Einstein's day, essentially no evidence one way or the other existed to support or refute his theory. Since 1905, however, extraordinarily precise experiments have been performed that confirm it.

I remember well the 1977 experiment of Alain Brillet and Jan Hall performed at the Joint Institute for Laboratory Astrophysics in Boulder, Colorado, where I was also working. Down in the lowest basement of the laboratory, they suspended a granite table, atop which was one of Jan Hall's stabilized helium-neon lasers. They rotated the granite slab and laser slowly, searching, as had Michelson and Morley, for an "ether wind" as the earth plowed its way through the hypothetical ether. The earth moves around the sun at about thirty thousand meters per second. Therefore, if an ether existed, one might expect to detect an ether wind of up to this speed. In the Brillet and Hall experiment, such a wind would have shown up as a small shift of the laser light's frequency of oscillation. Michelson and Morley's 1887 experiment had found no ether wind down to 4,700 meters per second. If it was smaller than this, their experiment could not have detected it. Brillet and Hall also saw no evidence for the ether, and their upper limit on the ether wind was fifteen meters per second, which is still the best Michelson-Morley style measurement to date.[16] Other kinds of measurements have lowered the highest possible value for the ether wind still further, down to 0.05 meter per second! As Einstein's first postulate implies, no trace of a luminiferous ether has yet been found.

Similar developments have taken place in tests of the second postulate—which says that the speed of light is independent of source motion. All manner of atomic light sources have been moved by experimentalists at speeds ranging from the very slow to ninety-two percent the speed of light. In every case the measured value of the speed of light was found to be the same.[17] Referenced to nothing (that is, without an ether), the speed of light is unvarying. If something comparable existed in spatial terms, it would have the property that no matter where you went, it was always the same distance away from you. You could get neither closer to it, nor farther away from it; the separation would be fixed. This is what infinity is like. No matter how far toward it or away from it one travels, it remains the *same* infinite distance away. As Einstein said, the speed of light plays the role of an infinitely great velocity. Light has no place, but it does have a speed and we are always separated from it by 299,792,458 meters per second.

Special relativity has been tested many times, and it has met all challenges. Einstein's run with light has truly transformed our understanding of it. Einstein had dared to ask what light would look like if one traveled beside it at the speed of light. The implications of the answer have been vast. We might well turn the question around and ask, what would the world itself look like while running with light? What is the world like from a "light's eye view"?

As every schoolchild now knows, nothing but light can reach the speed of light. No massive object, no matter how small its mass, can attain this ultimate speed. One would have to dematerialize to go as fast as light. Like young Einstein, imagine that you are light. Leaving your physical body behind, take refuge in your light-body and wing through space. What would you see?

It has become common knowledge since Einstein that as one's speed approaches that of the speed of light, space and time change in significant and surprising ways.[18] Lengths along the direction of motion shrink and moving clocks run slow. Exactly how these would change the appearance of things has been studied quite carefully in recent years, but I have never seen a serious treatment of what would happen if one traveled at the ultimate speed—the speed of light.

We do know that as one speeds up, the clocks we see flashing past us tick slower and slower, and the shapes of objects distort. Distances traveled seem to shorten—so-called length contraction. What happens when the impossible finally is achieved and we become light? Does time ultimately slow to a stop? To naturalists in 1911, Einstein remarked that what to us might be centuries would, to a living organism, be "a mere instant," provided the organism travels with nearly the speed of light.[19] What about space? Does distance completely disappear? In his seminal 1905 paper, Einstein declared that, "For $v = c$ all moving bodies—viewed from the 'resting' frame—shrivel up into plane figures." Everything is here and now, forever! In running with light, are we led full circle back to light eternal and omnipresent, outside of space and time? Or maybe Aristotle was right, and light is a universal and instantaneous change of state, at least when seen from the vantage point of light itself. By musing steadfastly on puzzles like these, perhaps we, like Einstein, will uncover yet another image of light.

—

EINSTEIN'S RELATIVITY, Planck's quantum of light, Steiner's angelic light, and Delaunay's artistic light all arose contemporaneously. Why? Remembering C. S. Lewis, we can ask: what were the questions most on the hearts of those living through the years from 1900 to 1925, and what do they reveal about the underlying tenor of their minds? During the previous three hundred years, every outer source of security had been removed. The guiding institutions of the past, whether sacred or secular, had been called into question or, especially in the course of war, destroyed. Simple faith no longer sufficed, existential questions loomed large, and few adequate answers were forthcoming.

Relativity theory, from one angle, can be seen as completing what was begun at the dawn of modern science by Copernicus and Galileo. In leaving the safety of sacred and classical traditions, we left much behind, but took with us still our body (space) and heartbeat (time). Though alone, we could cling fast to these. But no, according to relativity, even these were to be offered up. Space and time are not absolutely given. A more fluid structure and malleable rhythm is required. Thus every sure material prop, even the abstract comforts of absolute space and time, must be given over to the fire. We have faithfully followed the lineage of light to the twentieth century, and our place in the universe seems more completely meaningless than ever before. Many might deny it, but every external underpinning of classical science has been withdrawn. What, in the end, remains? Emerson's refrain, penned decades earlier, sounds clear and true: "Nothing is at last sacred but the integrity of our own mind."

The human spirit had, until the modern era, been cared for and sheltered from such disconcerting knowledge by church or state. Like the innocent youth Parzival, we had been raised in idyllic ignorance of such harsh conceptions of the world. Only when surprised by knights in glittering armor, whom he mistook for angels, did Parzival leave home dressed as a fool, without so much as a look back at his distraught mother. She died at his departure, and he had much to learn. Year after year, Parzival discovered his failures and stupidities, but never relinquished his aspiration. Finally, utterly alone in the dark and cold of the forest wilderness, a tattered figure, he dropped the reins on his horse's neck and, profoundly humbled, found his way back to the Grail castle as its new king. Like Parzival, light had first lived in an enchanted countryside as a favored and divine child. Then, donning the armor of theory, it rode forth to conquer all the phenomena of light. Early success bred overconfidence, but not all phenomena fell before its lance. Candlelight, the aurora, and lightning, to name but a few, stood firm. The old armor was stripped off and another far more subtle took its place, made of mercury instead of steel. The ramparts of the Grail castle vanished from Mont Salvat; they reappeared as the close-fitted plates of the human skull, the earthly sanctuary of the mind.

There, within that infinite interior space, Einstein created a theory possessing deep inner consistency, that urges a new self-reliance. With two postulates as anchor points, a brilliant thought structure arose that violated common sense, but fostered our courage to think devotedly. Einstein seemed endowed with a kind of inner gyroscope that allowed him to keep his bearings in every storm. I think his poise was but a reflection of his faithfulness to thinking, and to a view that held nature to be an expression of Intelligence. What *Lebensphilosophie* rejected was not rationality, but the lifeless form it took during the late nineteenth century. I would suggest that we see relativity theory as a reflection of an evolutionary movement of the human psyche toward true autonomy. The inflexible geometry of past science reflected the restricted structure of its imaginative base. By contrast the dynamic imagination of modern science liberates, but it also endangers. The stability we once found without, we must now find within. One of the most beautiful examples of this change occurred around the world's geometry. Not since Brunelleschi invented linear perspective had the geometry of human experience undergone such a radical shift.

In the decades leading up to the period of relativity theory, the very architecture of space was revolutionized. Until then the mathematical imagination, and with it all of scientific thinking, had been dominated by a single book. No text, other than the Bible, had done more to shape the thinking of the West than Euclid's *Elements*. Since the invention of printing alone, more than one thousand editions have appeared. Yet the mathematical framework the *Elements* espoused grants an unfounded privilege to one view, excluding the very idea of non-Euclidean geometries. The roots of a more flexible attitude to geometry reach back to the Renaissance creators of linear perspective, but the development of their first insights into the modern discipline of "projective geometry" had to await the work of great mathematicians such as Poncelet (1788–1867), Cayley (1821–95), and Klein (1849–1925). By the time of Einstein, non-Euclidean geometries and the even more comprehensive theory of projective geometry had broken the grip of Euclid on mathematical and spacial thinking, and a new imagination of space could be born.

To show something of the flavor offered by the new geometry, consider the circle and cube of Euclidean geometry. Once given, these forms are fixed. The curvature of a circle is absolutely uniform and determined by its particular radius; similarly, the corner angles of a cube are always ninety degrees and all sides are of equal length. In projective geometries these invariants disappear. The circle and cube can undergo an infinity of metamorphic changes. The figure below exemplifies but one of the possibilities; you must imagine the multitude of others not drawn.

Olive Whicher's lovely drawing allows us to see geometric forms as crystallizations born from light. Rays, like beacons, stream through space in an orderly fashion; their intersections defining and lengths embracing familiar geometrical figures. In addition, these rays must be imagined in motion, the figures ever-changing, so that all stasis disappears. The space of forms is entirely mobile, in flux, a geometry of streaming metamorphic life, and mathematics is alive with the vital force of the modern imagination.

Now freed of Euclidean constraints, we have to support ourselves on the gossamer wings of the geometry that our own hands have crafted. Seen in its light, the contours of continents we overfly may change strangely from one moment to the next, but

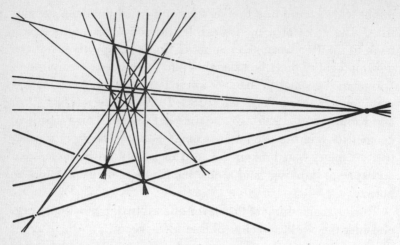

A projective geometric construction.

as Blake once wrote, "No bird soars too high if he soars with his own wings."

Light and the Architecture of Space-Time

Structure is the giver of light.
Louis Kahn

The new architecture of space presupposes architects to imagine it, and builders to build it. Like a modern gnostic, the architect Louis Kahn was constantly attentive to light.[20] Where silence and light met, in the space formed by their union, he discovered inspiration. Here was the Sanctuary of Art, the Treasury of the Shadows. Within this sanctuary, he wrote, "the artist offers his work to his art."

Silence to Light
Light to Silence
The threshold of their crossing
 is the Singularity
 is Inspiration
(where the desire to express meets the possible)
 is the Sanctuary of Art
 is the Treasury of the Shadows
(Material casts shadows, shadows belong to light).

Silence broods. In it resides the restless, "unmeasurable desire to be." When the desires of Silence meet Light, "the measurable, giver of all presence," worlds arise. From these two originates all that is made.

> ... all material in nature, the mountains and the streams and the air and we, are made of Light which has been spent, and this crumpled mass called material casts a shadow, and the shadow belongs to Light.
> So light is really the source of all being. And I said to myself, when the world was an ooze without any kind of shape or direction, the ooze was completely infiltrated with the desire to express, which was a great congealment of Joy, and desire was a solid front to make sight possible.

To make sight possible. Around and within that original congealment of Joy rayed Light, the source of all being. Spending itself, it built up not only mountains, streams, and air, but also the sighted creatures who see it. As Richard Wilbur writes, "It was the sun that bored these two blue holes," our eyes. Kahn's cosmogony was one of light, the world its offspring.

Architecture and the new geometry of space-time faced each other in the first light of the new century, at once intellectually liberated and spiritually inspired. Like figures shaken from a textbook of projective geometry, buildings of glass, steel, and concrete grew in the landscape. The organic architecture of

Gaudi in Barcelona, Steiner in Switzerland, Le Corbusier in France, caught in concrete the dynamic morphology of organic forms. Bruno Taut's "Glass House" allowed light to magically penetrate even its structural walls through glass bricks. To invert a thought of Kahn's, light seems the giver of space.

THE INTERPLAY OF light and space displayed yet another facet in the hands of Einstein when he extended his special theory of relativity to embrace the phenomenon of gravitation. His general theory originated with what Einstein called the happiest thought of his life.[21]

—

IN NOVEMBER OF 1907, while sitting in a chair in the Berne patent office where he worked, it suddenly occurred to Einstein, "If a person falls freely he will not feel his own weight." A simple observation, but within it lay the gravitational analogue of Faraday's archetypal experiment of electromagnetic induction. The relativity of electric and magnetic fields that had led Einstein to his special theory here possessed an exact parallel in gravitation, one that would lead him ultimately to his general theory of relativity. Einstein's "happiest thought" has since become enshrined as the "equivalence principle" of relativity theory. It states that gravitational attraction is experimentally indistinguishable from steady acceleration. Confined to a room, the pressure on your feet could as well be due to it accelerating upward (as in an elevator or spaceship at blast-off), as to the gravitational attraction of the earth.

Einstein was able to use the equivalence principle to predict that the path followed by light would be bent as it passed a massive body such as the sun. Arthur Eddington's 1919 English expedition to Principe Island off the coast of Spanish Guinea succeeded in capturing in photographs just this effect. The deflection was tiny, but in essential agreement with Einstein's

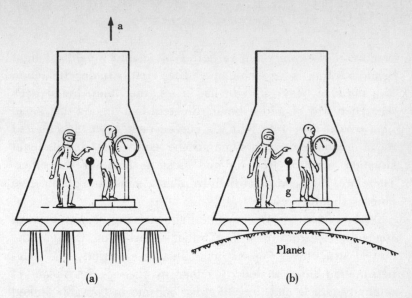

(a)   (b)

Einstein's principle of equivalence. To the astronauts inside, the experience of acceleration is exactly like being at rest on a planet.

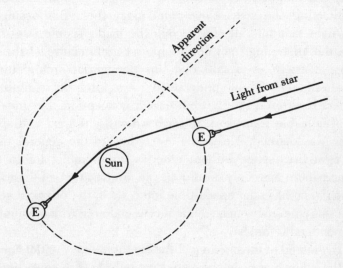

The bending of starlight by the mass of the sun, as predicted by Einstein. The apparent change in a star's position was first demonstrated by Eddington's expedition.

prediction of seven years earlier. Why should weightless light bend as it travels through empty space simply under the action of a massive body? If light had mass, then fine, but it does not. Then why should it bend? Einstein had argued that since light would appear to follow a curved path in an accelerated room it would do the same in the equivalent gravitational situation; what is true for one must be true for the other. Therefore, the bending of light by planet, suns, and all massive objects should occur.

No longer should we imagine light piercing its way through space in straight lines, but rather as weaving through the curved structure of "spacetime." Just as electricity and magnetism had been wedded together by the special theory of relativity, space and time likewise became inseparably linked in the reality of Einstein's universe. The pathways of light held a special place in the new theory, for while they curved, they still mapped out the shortest distances or "null geodesics" connecting one space-time point to another. If the sun and starlight that fills all space could be made visible, we would see the glistening, mobile, sculptural architecture of the cosmos. Einstein even said that this immaterial architecture of space-time could be understood as a kind of rehabilitated ether. It differed utterly from earlier conceptions but provided the malleable framework through which light sped and space-time was defined. He said, "According to the general theory of relativity, space without ether is unthinkable; for in such space there not only would be no propagation of light, but also . . . no basis for space-time intervals in the physical sense. But this ether may not be thought of as endowed with the qualities of ponderable media. . . ."[22]

Dismissed at the opening of the century, a new understanding of the ether has been suggested not only by Einstein, but also by prominent physicists since. Their proposals are based not only on relativity theory but also on their attempts to understand quantum theory. The material ether of Kelvin is dead, but for some the ether lives anew in a far more subtle, immaterial guise.

—

DURING THE FIRST decades of the twentieth century, Einstein's meditations on light never flagged. Nearly everyone who visited him in later years seems to tell the same story. He had done much to father the strange quantum of light, but he was convinced that quantum theory could not be the entire story. Many were the revolutions of the first decades, but it seemed to him that the demands of causality required a nonprobabilistic theory. As he often said, "God does not play dice." To Einstein the quantum physics of the microworld was still incomplete, a fragment of the truth. We might think we have reached the end, but we are only fooling ourselves. Even the concept Einstein had introduced, the quantum of light or photon, could not be understood properly in quantum theory. In 1951 he declared, "All the fifty years of conscious brooding have brought me no closer to the answer to the question, 'What are light quanta?' Of course today every rascal thinks he knows the answer, but he is deluding himself."[23]

It is forty years since Einstein warned against the arrogance of scientific surety regarding light. Efforts to understand light have not abated since his death, and yet the essence of light remains an enigma.

—

WE STILL LACK a picture for the photon. A remarkable feature of relativity is that its predictions, including those it makes about light, do not rely in any way on a model for light. Relativity does not need to know whether light is a wave or particle, it does not need to answer the question "What are light quanta?" in order to predict the bending of light around the sun. Whatever light is, it must conform to the principle of relativity. Much of relativity's beauty and universality stem from this remarkable feature of the theory, one which Einstein appreciated deeply. How far can one go without a picture of light, how many predictions can one make without knowing what light really is?

In recent years, this avenue of thinking has been pursued, especially by the remarkable late physicist Richard Feynman. Following the path of beauty and not that of image, Feynman stalked the phenomena of physics with an ancient and honorable instrument, the confidence that perfection lay at the root of existence. The dictates of beauty have commanded many over the centuries, guiding the imagination of inquirers down to the present. From the shapely attractions of space-time we turn to the abstract beauty of perfection.

Following the Beautiful

It isn't that a particle takes the path of least action but that it smells all the paths in the neighborhood and chooses the one that has the least action.

Richard Feynman[24]

Something of a legend in his own lifetime, Richard Feynman was a brilliant young prankster at the secret Los Alamos laboratories where J. R. Oppenheimer guided his scientific team to produce the first nuclear bomb. As a provocative member of the congressional committee to investigate the *Challenger* disaster, Feynman's irrepressible genius revealed the tragic flaw of booster design responsible for the mishap.

For someone of such brilliance and originality as Feynman, what was it about physics that excited him and drew him into the subject originally, then sustained his fascination with it for decades? When all the hard things come easy, as they did for Feynman, to what does one hold fast? The answer lay in a remark made to him by his perceptive high-school physics teacher, Mr. Bader. Noticing the boredom of his exceptional student, he gave him something lovely to think about. In the vocabulary of physics, Mr. Bader told Feynman that light, like all things, always follows the path of the beautiful.

Bending Light, Again

While fishing, every child has noticed that when you dip a fishing pole into the water, it *seems* to break. Put a spoon into a glass of water, and when seen from the top, the spoon shows a kink in it where none appeared before. We are not surprised that when removed from the water, both the fishing pole and spoon regain their pristine forms. We are accustomed to such behavior in spoons and fishing rods. Perhaps we even know something about refraction; that light (or the visual ray) bends when it passes from one medium to another: from air into water, for instance.

Why does light do this? Why does it bend when entering water from the air? Or, to enlarge the question, why does it travel just as it does in general, sometimes bending just so much, other times reflecting at exactly a certain angle? Similarly, why does a stone or an electron, or anything, travel in exactly the way it does? What stands behind the behavior of all physical systems, behind all the laws of physics? It seems hugely presumptuous even to ask such a grand question, and yet at a very powerful level, there is a single fascinating answer. It was about this that Mr. Bader spoke to the young Feynman that day, and I believe it was this that held Feynman for a lifetime.

The actions of light can once again lead us to the heart of

things. By puzzling with others over the simple bending and reflecting of light, we will be led in the end to Bader, Feynman, and, surprisingly, to the contemporary challenge of quantum mechanics. The tale begins in the exotic location of second-century Egypt.

—

THE FIRST SYSTEMATIC observations of refraction seem to have been made by the Alexandrian astronomer Claudius Ptolemy in the second century A.D. From his important treatise on optics there has survived what appears to be a table of data giving the angles of light in air and water.

ANGLE IN AIR	ANGLE IN WATER
10	8
20	15½
30	22½
40	29
50	35
60	40½
70	45½
80	50

At first blush, the numbers all seem reasonable. The angles in water are all less than the angles in air as they should be; the overall pattern seems right. One becomes suspicious, however, when one considers the reported accuracy of the measured angles in water. To measure to half a degree of arc is pretty remarkable for the second century A.D. Look more closely at these angles, and subtract adjacent numbers from one another. If you do so, the following results are obtained.

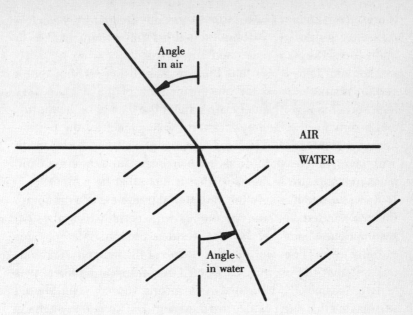

Refraction at an air-water surface.

ANGLE IN AIR	ANGLE IN WATER	DIFFERENCES
10	8	7½
20	15½	7
30	22½	6½
40	29	6
50	35	5½
60	40½	5
70	45½	4½
80	50	

The column of differences is clearly an arithmetic progression: 4½, 5, 5½ ... and so on, adding a half each time. Careful comparison with angles in water as measured today shows that the actual angles are not as given by Ptolemy at all. In other words, the column of water angles was not created by measurements but by theory. Ptolemy forced his "measurements" to fit his theory. They are too "good" to be true.

The point here is *not* that Ptolemy was dishonest, but that a certain motive existed for his falsification of the data, a very high sort of motive. Namely, he wanted the data to be beautiful, and beauty for the Hellenistic world was geometric, the perfection of mathematics. The angles in water just *had* to be in an arithmetic progression. It is such a beautiful sequence, and measurements are so inexact, that it just must be correct.

As was also the case in his magisterial treatment of astronomy, Ptolemy started his study of optics with certain metaphysical assumptions about the way the world is ordered. Metaphysics literally means "beyond physics." One of his metaphysical values was that the universe is formed according to number; it is truly a "cosmos," which in Greek means "order." Nor should we imagine that such values are outdated. Einstein, among many others, tenaciously upheld the values of beauty and order as essential features important to the theoretician. Nor was Ptolemy alone in his own day.

Although he disagreed with Ptolemy about the specifics of refraction, the fourth-century philosopher Damianos displayed a similar spirit when discussing refraction in his own treatise on optics. He derived the law of refraction on the basis of the following logic: "if Nature does not wish to permit our visual ray to wander about fruitlessly, she will let it break at equal angles." Thus, according to Damianos, there is a single "best" way for nature to refract light, otherwise fruitless visual chaos would result. The best way is to arrange for the angle of refraction always to be exactly half of the incident angle.

Eight hundred years later, Grosseteste agreed with Damianos, making even clearer the metaphysical basis for the derivation. "And it is shown to us by this principle of natural philosophy, that every operation of nature is by the most finite, most ordered,

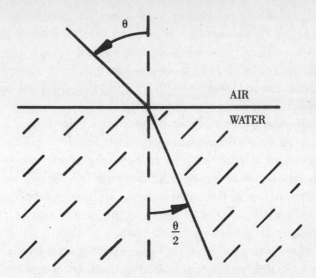

Grosseteste's early and erroneous law of refraction, based on his sense of the beautiful.

shortest and best means possible."[25] Unfortunately, the facts of refraction do not agree with the best possible world as imagined by Damianos and Grosseteste; shortest may not be best.

The gut scientific reaction to all this is to call such metaphysical argumentation nonsense. Yet Mr. Bader's conversation with Feynman sounded not unlike the musings of Damianos and Grosseteste, with an essential difference; Mr. Bader's definition of "best" was right! It was, in Plato's technical terminology, divine.

Causes: Necessary and Divine

Plato and, until the seventeenth century, all philosophers after him, held that causes were of at least two types; one he called "divine," while all others were merely "necessary." Divine causes are "endowed with mind and are the workers of things good and fair."[26] By contrast, necessary causes proceed by brute force and, left to their own devices, would lead only to chance, chaotic results if they were not directed by divine causes.

Carpenters with tools and materials at the ready, like necessary causes, still require an idea, usually expressed by architect and owner, before a house—the right house—can be built. As with house construction, so also with the world. The forces of nature in themselves are unintelligent, brute, accessory causes called forward to "execute the best as far as possible" by higher divine causes, says Plato.

Aristotle termed Plato's divine cause *causa finalis*, or final cause, distinguishing it from three other accessory causes: material cause (the lumber of the house), efficient cause (the builders with their tools), and the formal cause (the house plan of the architect). The final cause of the house is its use as a shelter. The end to which it is put, its purpose, is the *causa finalis*.

In what possible sense can we make use of final causes in optics? When puzzling over the phenomena of refraction and reflection, one quite naturally asks after the forces at work, the pushes and pulls that act on the light (whether thought of as particles or waves). We do not ask what is the "best" path for light to follow. Yet there is a tradition in physical science—of which Damianos, Grosseteste, and Feynman are a part—that does ask just this question.

That tradition goes back at least as far as Hero of Alexandria (c. 125 B.C.), who proved that the path actually taken by the visual ray (or by light) in reflection from a mirror is the shortest possible path connecting the eye to the source of light with a reflection off the mirror. All other paths are longer. Light follows the best path, if by best we understand shortest.

Both Damianos and Grosseteste knew of Hero's derivation, and sought to adapt its argument to the case of refraction. The difficulty is that for refraction the distance traveled is patently *not* the shortest physical distance connecting light source and eye. So they argued for an alternative notion of best, namely equal angles. Where they failed was not in their use of metaphysical criteria, but rather in their choice of metaphysics. They needed one closer to Hero's criterion. They needed the notion of shortest *optical distance*, which usually differs from a straight-line distance.

Instead of the simple criterion of shortest distance, Fermat advanced his "principle of least time." Light *always* travels the path that minimizes not the distance, but the *time* it takes to go from one place to another. Like a savvy commuter trying to get to work during rush hour, light may well choose to go a longer physical distance if its travel time is shorter. The bending of light as it passes from air into water, refraction, is a case of light following a path that is longer in length but actually shorter in time. With Fermat's principle of least time, and no more geometrical knowledge than Hero possessed, it is possible to derive the true law of refraction (Snell's law), as well as of reflection. Light does follow the best path, if only we know what is meant by best.

Traveling All Roads

If the story ended here, it would already be wonderful, but Fermat is only the first modern to apply a much broader and more powerful, unifying principle that is much loved in contemporary physics. What Mr. Bader told young Feynman was the twentieth-century version of Fermat's principle. Generalized by the French mathematician Maupertius in 1744, it is known today as the "principle of least action." The specifics are not important, but suffice it to say that for an astonishingly wide range of physical phenomena, be it optics, mechanics, hydrodynamics, electrodynamics, or quantum theory, one can define a quantity called "action," and go on to derive the laws of physics for the domain of interest by finding the path along which the action is smallest.[27] Thus its name: the principle of least action.

The father of the modern quantum theory of radiation, Max Planck, saw in this principle a very high truth. Unlike the "differential" laws of physics that specify the moment-by-moment forces acting on a particle as it moves, the principle of least action seems to work with the whole path, undivided. Formally, one can always pass from the integral to the differential formulation, but each carries with it a particular attitude toward the world. In one—the differential view—we are the general contractor overseeing a construction crew and the materials used; we are concerned only with necessary causes. But we can also step back from our manual labors to admire the work even before it has reached its physical completion, seeing it whole in our mind's eye. In doing so, we are free to imagine other designs, other paths chosen, and then to select the best, to conceive a cause divine. Both ways of seeing the task are valid, each has its rightful domain. In physics, the judgment is nature's, she has chosen her "divine cause" and it seems to be the principle of least action.

What excited the young Feynman was, I believe, the experience of seeing the beauty and wholeness of this universe, of seeing into the choices that nature had made. That first seeing, which was an instance of scientific inspiration, bore splendid fruits when later, as a brilliant graduate student at Princeton, Feynman created his "path integral formulation" of quantum theory on just this same principle of least action. It has been called by many the most beautiful of all formulations of quantum theory. In its view, all roads are taken, all paths are traveled by a quantum particle, be it light or matter. When taken together the countless paths distill to one whose action is least. Whatever light is, here is where we will find it.

—

TWO AVENUES OF research lead us into light. One is directed toward the universally true, and operates with mighty principles such as relativity and least action. The other tends toward the infinitely small, the building blocks of the world. Although we call them by other names, even today we allow Plato's distinction between "divine" and "necessary" causes. Einstein termed them principle theories and constructive theories.[28] With the former we flatter ourselves that we have detected the rationality of the universe, its underlying intelligence; with the latter we are concerned with the nitty-gritty engineering of how nature realized her objectives. The play being written, we are curious to peer behind the wings at the stage machinery. Goethe and most of twentieth-century philosophy warn us not to make too much of the "reality" we discover there. The discoveries we make, the "realities" we see, may well reflect more about us than the object of our investigation, and this has proven doubly true of modern physics. Yet one is loath to forgo the pleasure of drawing pictures from our theories. We wish to know not only how light behaves, but what it is. Unwilling to accept the answers of the past, we forge our own understanding, cast our own idols.

Neither the theory of relativity nor the principle of least action speaks about the nature of light. Starting from apparently innocent beginnings with intuitively sensible assumptions, however, these theories do make extraordinary deductions that lead to universal laws of beauty and power. Yet for some, laws alone are not enough. Another avenue must also be followed, and it is the path of dissection. We know it from previous chapters. Ever since Leucippus and Democritus inaugurated atomist philosophy, science has sought for the elementary constituents of nature. Light will of necessity be reduced through this analysis to its least parts. And yet here, too, a thousand surprises await, because the least part of light quite literally conceals a whole.

We live poised between the part and the whole, the differential and the integral, the fragmentation of a darkly ahrimanic spirit and the enticements of a luminous luciferic one, between, as Schiller called them, the tendency to substance and the tendency toward form.[29] Between them both we find our way, giving to each its due. Schiller likened this kind of engagement with the world to true play, for only in play is one free.

—

LOUIS KAHN ONCE told of a task he imposed upon himself.

> I gave myself an assignment: to draw a picture that demonstrates light. Now if you give yourself such an assignment, the first thing you do is escape somewhere, because it is impossible to do. You say that the white piece of paper is the illustration; what else is there to do? But when I put a stroke of ink on the paper, I realized that the black was where the light was not, and then I could really make a drawing, because I could be discerning as to where the light was not, which was where I put the black. Then the picture became absolutely luminous.

Even the artist must invoke darkness in order to illustrate light. Louis Kahn knew this truth more clearly than most. I am no different. The finite expressions of language compose my palette, and my limited conceptual world defines the dim markings I can make. Like Kahn, all I am able to do is put my words where light is not. But perhaps by my doing so, this book, too, will become absolutely luminous.

Silence and Light by Louis Kahn.

—

CHAPTER 11

Least Light:
A Contemporary View

Ever splitting the light! How often do they
strive to divide that which, despite every-
thing, would always remain single and whole.

Goethe

Our discussions have reached a turning point. The changing
understandings of the past must give way to the tentative grop-
ings of the present. The stories told are now our stories. In our
best judgment, formed on the basis of everything we know, what
do we regard as the essential nature of light? To rainbow, candle,
prism, and mirror are now joined deeply puzzling quantum phenomena in which the character of light seems paradoxical in the
extreme.

Rather than describe this "modern" light abstractly, we will
search among the countless experiments of quantum optics for
those concrete instances that most clearly express the photon's
essential behaviors. These experiments will dramatize the
unique features of light quanta, and become what Goethe would
gladly have called the archetypal experiments of the quantum

domain. By patiently working with them, we will, in Goethe's view, fashion new organs of cognition better suited to our subject than the traditional faculties we have inherited from classical physics. The challenge of quantum phenomena is, then, a challenge to our complacency of mind, an incentive to reach beyond ourselves. Like fish swimming the black waters of Mammoth Caves, we have become attuned to the darkness. Only after centuries of effort have we moved into light-filled waters. Blinded still by the habits of the cave, we are reluctant to explore and fully contemplate the open sunlit terrain. Only by self-consciously doing so can we ever hope to approach an understanding of light.

The phenomena of light, even when restricted to quantum effects, are innumerable. Yet certain of them teach us more than others. Like a great sculpture or work of art, in them the essential is brought to the surface, and every extraneous feature is chiseled away. Such phenomena carry a special significance and, in each field, are few in number. Goethe called them "archetypal phenomena," and I will adopt his term here as well. In them we encounter self-evident manifestations of natural law. They are the experiential windows that look into the eternal principle by which nature shapes the flow of phenomena. If we have the eyes really to see an archetypal phenomenon, then we have the wit to understand it. What are the archetypal phenomena of light at the quantum level?

By turning away from the common experiences of light to the quantum, we follow the modern trend toward a study of the least parts of light. Contemporary quantum experiments investigate light when reduced to its weakest possible level. As we saw, Max Planck and Albert Einstein were the first to posit the existence of an elementary quantum of light, suggesting light was not infinitely divisible, and so possessed a least part. In 1926 the American chemist G. N. Lewis gave the quantum of light its current name: the photon. Research physicists have probed the photon and quantum optics is the specific subfield that takes upon itself the study of light as a quantum object. As I have grown to appreciate firsthand, it is a marvelous study full of dramatic moments that ultimately offer a fresh way of thinking about light.

Can we find archetypal experiments in which the photon clearly, if paradoxically, shows some aspect of itself? Then, like Einstein steadfastly running after light, we can attend single-mindedly to the puzzling features of the photon, hoping that by doing so we might finally come to understand the mystery it presents.

If we wish to study the least part of light, the isolated photon, the first task clearly is to construct a source of them. At first blush it would seem trivial. Simply take any light source, a candle even, and dim it sufficiently so it gives off a slow but steady stream of single photons. Just this was done for decades until the fundamental logical flaw in doing so was realized. No one ever checked experimentally to see if these sources were really producing single photons. One assumed without experimental evidence that dim light is a single-photon stream. What evidence would convince us that a source is a single-photon source? Obviously, a suitable means of testing light sources is required.

Perhaps the most elegant single-photon test has been realized by the French group of Alain Aspect, Philippe Grangier, and G. Roger.[1] Although technically somewhat difficult, it is conceptually supremely simple. One passes the suspected photon into an optical instrument that divides light, sending half one way and half another way. If at some point the light is no longer divisible, then we have reached the level of the photon, or "atom" (from the Greek, meaning indivisible) of light. The optical device that divides light in half is a half-silvered mirror, or "beamsplitter." When light falls on it, half the light is transmitted and half reflected. Imagine a single, atomlike photon

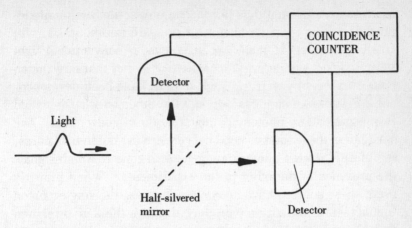

Light as a particle. In this experiment a single photon falling on the half-silvered mirror goes to one detector or the other, never both.

striking the beamsplitter. What will happen? If it is truly indivisible, then it will go one way or the other but *not* both. This is called "anticorrelation." If, on the other hand, the light can be divided, both branches will have light, half will go one way and half the other. This is the test.

In contrast to an atomic theory of light, the wave theory conceives of light as infinitely divisible; there is no lower limit to how weak the light's intensity can be. Therefore the beamsplitter will *always* divide the light, transmitting half one way and reflecting the other half. The beamsplitter is, therefore, a litmus test for light: wave or particle. Performing it on a variety of light sources has yielded startling results. All traditional sources such as candles, incandescent lamps, gas discharge or fluorescent lights, and even lasers show no anticorrelation, whatever their intensity. No matter how weak their light, the data show that the light they emit splits at the beamsplitter in a way that can be well described by a wave theory of light. All the usual sources of light thus fail the simplest criterion for single-photon emission. Single-photon sources are not natural.

All is not lost. Humans are ingenious. If conventional light sources do not suffice, we can contrive new ones that serve much better. In recent years, two such sources have been developed, one relying on atomic cascade and the other on what is called in the trade "two-photon, parametric, down-conversion." The specifics of these sources need not concern us, but in both cases, two closely related photons are generated, one of which signals the presence of the other to the experimenter. When properly used, these sources have been shown to pass the single-photon litmus test. The light they produce displays the anticorrelation looked for in single-photon sources.

Thus, only in recent years have physicists had good sources of nonclassical, quantum-mechanical light, and we have been enjoying ourselves ever since. Dozens of elegant experiments, often yielding strange results, have been done in recent years using single-photon sources. I will relate a few that I think most clearly reveal the subtle nature of light. Within them we may find the archetypal experiments we are seeking. Now that we have the single photon, let us study its behavior. It may be hopeless to express the nature of light abstractly, but through its actions we can catch a glimpse of its character.

Archetypal Instances

Behind the tireless efforts of the investigator there lurks a stronger, more mysterious drive: it is existence and reality that one wishes to comprehend.[2]

Einstein, 1934

Early on, Einstein drew attention to the perverse nature of the photon. As usual, he suggested an insightful thought experi-

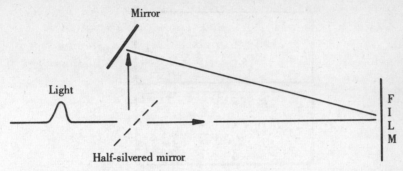

Light as a wave. Interference requires that light travel both paths to the film.

ment. Use a single-photon source, he once said to Bohr, to generate an isolated photon. Pass the photon through a beam-splitter. We *know* that the photon has gone one way or the other; this was after all the very test we used to declare the source fit to be called a single-photon source. Check as often as you like; the anticorrelation is always there. Now, said Einstein, replace the two detectors with mirrors that redirect the photon so that whatever way it travels, it ultimately lands on the same section of a photographic film. What do you see?

With the arrival of the first photon, a single, tiny black spot appears roughly where one expects. The following few photons likewise show up as spots scattered around the first. So far everything is fine. But as time passes and the spots accumulate one by one, something altogether unexpected appears. The spots align themselves into dark and light bands, into clear interference fringes. The experimental facts are unambiguous. When we set up to see interference fringes, we see them, even with single photons. When we set up to see which single path the photon has taken, we find out. The problem is not with the phenomena, but with the inadequate thoughts we bring to them.

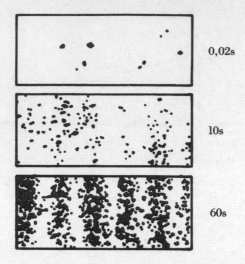

0,02s

10s

60s

Interference fringes build up slowly, one photon at a time.

In the first case, we think waves; in the second, particles. In the first, we think both paths; in the second, one path. How can both be true? What kind of a thing could possibly behave this way? To quote Einstein at sixteen, "something like that does not seem to exist"! Hold on to this thought for a moment.

Ever since Thomas Young, interference fringes have been understood as due to light traveling by *two* paths. But the source of light used in this experiment was from a single-photon source, which has the demonstrated property that it sends photons along only *one* path, not two. The appearance of interference fringes, however, requires that in some sense the single, indivisible quantum of light, the lone photon, takes, or at least is affected by, both paths! Bohr paraphrased Einstein's summary of the situation as follows: "In any attempt of a pictorial representation of the behavior of the photon we would thus meet with the difficulty: to be obliged to say, on the one hand, that the photon always chooses *one* of the two ways and, on the other hand, that it behaves as if it had passed both ways."[3]

Here is an archetypal instance of wave-particle duality. For several decades, Einstein's thought experiment lacked a proper laboratory realization. Finally, in 1986, Aspect, Grangier, and Roger successfully performed a lovely version of the experiment with unambiguous results. The solitary photon does indeed interfere with itself. One thing—the photon—seems somehow simultaneously related to two distinct paths. The implications of the experiment are fundamental. The root structure of quantum mechanics, and our idea of light, must accommodate this experimental fact. We have found an archetypal phenomenon but lack the ideas with which to see it rightly.

Goethe was right. Try though we may to split light into fundamental atomic pieces, it remains whole to the end. Our very notion of what it means to be elementary is challenged. Until now we have equated smallest with most fundamental. Perhaps for light, at least, the most fundamental feature is not to be found in smallness, but rather in wholeness, its incorrigible capacity to be one and many, particle and wave, a single thing with the universe inside.

In recent talks and articles, the distinguished physicist John Archibald Wheeler has reminded the physics community that the quantum ambiguity demonstrated by single-photon interference experiments is even more serious than it first appears. He did so by proposing an interesting thought experiment, one first hinted at by Einstein and Carl Frederich von Weizsäcker.[4] In it the previous two dichotomous experiments (anticorrelation and interference) coexist until the bitter end, and the wave-particle ambiguity is sustained throughout the photon's flight.

Imagine that a single photon passes through a beamsplitter as before, but the choice is delayed as to whether detectors or mirrors are installed. In principle, photons can be "split" at huge distances from the detectors, and the delay made as long as one pleases, years even. The puzzling nature of light sinks in while we wait. During that entire time, what can we say about the path followed by the photon? If we think of it as a particle, then, when the paths split long ago, the photon took one of them. However, we may well decide to run an interference experiment on that photon, and make our decision to do so well after the photon has passed the fateful decision point. Nevertheless, we will see interference. If, on the other hand, we think of the photon as a wave going both paths, then again, long after the branch point has been reached, the experimenter can switch to a measurement of path, again with success. Prior to the final moment of measurement, it seems that we cannot say the photon travels both paths, but nor can we say it travels one path; a situation that is certainly very awkward! However, after we complete the measurement, we *can* apparently declare whether the photon had traveled to us by a single path or by both. In Wheeler's words, by our delayed choice we seem to "have an unavoidable effect on what we have a right to say about the *already* past history of that photon." Delaying the choice of what experiment we run has dramatized the essential ambiguity of single-photon physics.

Something very strange confronts us here. Conventional, well-defined objects have a clear, unique history. They originate somewhere, travel a path, and in the end arrive somewhere else. Single photons (indeed, all quanta) are also well defined, at least mathematically, but are different. The photon's history of motion shows an irreducible and paradoxical, wave-particle character that is unlike anything we know from classical physics. What is the source of the ambiguity?

At the Max Planck Institute for Quantum Optics outside Munich, I recently collaborated in a demonstration of Wheeler's proposed delayed-choice experiment.[5] The results were as quantum theory predicted, convincingly showing us that a single photon is ambiguous in the quantum-mechanical sense all the way until the detection. The single photon makes no "decision" at the beamsplitter, either for particle or for wave. It is neither, or both. The ambiguity is essential to light, one exceedingly difficult to grasp but all the more important for its elusiveness. If the photon exists between sightings, then our description of that existence must allow for the facts of single-photon interference experiments.

—

ATTITUDES TOWARD THE delayed-choice experiment, and other experiments like it, vary. The majority of physicists by far simply do not concern themselves with the meaning of their quantum calculations. Nor do they trouble themselves about the implications of archetypal quantum experiments. Science is not, they say, concerned with truth or meaning, but only with prediction and control; it is an instrument. Nineteenth-century scientific arrogance here changes to twentieth-century cynicism. Power we have and that is enough; true knowledge we relinquish forever. According to this view, modern science is like the astronomy of ancient Babylonia three thousand years ago. Without any physical notions about our solar system, astronomer-priests could predict the course of stars, sun, moon, and planets with amazing accuracy by applying purely arithmetic procedures to data from the past. Likewise, we can apply a quantum algorithm to make predictions about quantum phenomena without any notion of what a photon, an electron, etc., really is. Yet there is a difference. Babylonian astronomer-priests sought relentlessly for the meaning of their observations, but it was a spiritual meaning wedded to rituals, practical life, and their participation in the life of their gods and goddesses. With this they were content. We have shed every spiritual framework for science. The instrumentalist view would also abdicate hope for a true physical understanding.

As the gods disappeared the Greeks fashioned a rational,

geometric cosmos to replace the dying spiritual one. In the sixteenth century, matter joined reason, and the universe gradually became a clockwork. Twentieth-century physics and philosophy called much of this view into question. It asked whether any true knowledge of matter, light, or by extension, all of nature was possible. The deconstruction of scientific knowledge has been liberating, but freed from its tyranny, we now verge on a demoralizing nihilism. If scientific knowledge is only instrumental, is there meaningful knowledge of any kind? In their hearts, no serious, practicing scientists believe in instrumentalism; no teacher glows warm with enthusiasm in presenting a meaningless calculational procedure to his or her students. To what, then, does physics lead?

When we are confronted with the black void of pure instrumentalism, the temptations of reactionary idolatry are very near. Having lost the gods, we fall in love with the beautiful idols we can raise in their places. Atoms, quarks, tiny black holes . . . they are reified, garlanded, and dragged forward to assume a place in the temple. Calling them real, we animate them with the false life of fear, our fear of the unknown being of nature. We feel caught between the sophisticated emptiness of deconstructionism and the gaudy creations of our own hands. To what can one hold? Goethe's answer would have been that given to Hegel and the readers of his *Theory of Color;* hold to the phenomena. They can be trusted. Rightly seen, they will become the theory.

To see we must represent objects to ourselves, we must apply to them concepts that fit their nature. Failing this, we fail to see. We are like the once-blind S.B. who remained mostly sightless even when surgery repaired his eyes. He possessed the raw material for sight, but did not see. The experiments of quantum optics have given us the raw material of light, but we still lack the concept suited to its nature. Once we fully possess it, the paradox of wave-particle duality will disappear while "wave-particle-ness" remains. When that intelligence is brought to archetypal quantum phenomena, we will see them with understanding and not confusion. In Emerson's language, we will have "named" them, and once named, a phenomenon becomes a theory. We can then recognize the presence of "wave-particle-ness" in many places, having seen it clearly in one. Like tenderness discovered in childhood, we will know it when we meet it.

—

LIKE ADOLESCENTS READING Shakespeare, we feel the language of quantum phenomena to be at once strange and wonderful. The meanings are obscure, some words unknown. All the more reason to pause over it, to reread the text and muse over its possible meaning. Understanding it to be a play, we can mechanically act out the parts, but much is left untouched thereby. If only the script could be penetrated, the meanings discerned, then how much richer the production would be. Surely nature, like a found dramatic script, deserves to be read for meaning and not merely acted out for profit. As we hear the Sirens of the past, we should know that everything depends on the courage and freshness of the reader. The phenomena of light refute a nineteenth-century reading of the world, even if the modern staging remains fragmentary and hard.

In recent years growing dissatisfaction with instrumentalism has led many scientists to adopt other attitudes toward quantum phenomena. Some espouse "critical realist" perspectives, including realist interpretations of quantum mechanics. Regarding quantum phenomena, there are, broadly speaking, two such positions. The larger group holds that we must fundamentally change our understanding of the photon and all other elementary particles. They are real but profoundly unfamiliar. For example, one cannot speak about the "path" of the photon. The photon does not have a trajectory as such, a set of particular positions that change with time. In order to describe the photon's history we must rise above the particulars to a more abstract level. One adds together the "quantum amplitudes" for all possible paths the photon can travel, and the "superposition state" so constructed is the proper quantum mechanical description of the photon. The evolution of the superposition state replaces the concept of trajectory familiar to us from classical physics. If one takes a realist view, then the superposition state, which is a kind of metaphysical mixture of possible paths, *is* the photon as it travels from beamsplitter to detector. Wheeler calls the superposition state the "Great Smoky Dragon." Although the head and tail of the dragon show up, its huge body is obscured by smoke. So too with the photon. We may know all about its emission and detection, but the path it follows in between is ultimately ambiguous.

Wheeler goes further, reaching beyond the views held by most physicists when he uses delayed choice to show how strongly our present measurements connect us with the past. In the above experiments, only the final act of detection lifts the quantum ambiguity and dispels the smoke to reveal a classical world in which even the photon's path can be talked about. The final detection apparently determines the already-past history of the photon. In an interview, Wheeler remarked, "To the extent that it [the Great Smoky Dragon or photon] forms a part of what we call reality, we have to say that we ourselves have an undeniable part in shaping what we have always called the past."[6]

A second and much smaller group interprets the single-photon, delayed-choice, interference experiment very differently. David Bohm and Basil Hiley are the foremost advocates of a view that pictures photons and all quantum objects in more traditional ways. Quanta have positions, trajectories, and real path histories. Of course Bohm and Hiley must provide in their theory for the paradoxical phenomena of quantum mechanics. They do so by introducing something unknown until now. Physics has traditionally divided the world into particles (electrons, quarks ...) and fields (gravitation, electromagnetic, and nuclear). They suggest the addition of a third entity, the so-called quantum potential. It acts like an immaterial ether whose purpose is to guide the photon in something like the way a remote-controlled airplane is guided by radio signals. All quantum effects are carried by this information field.

The quantum potential has unusual features. First, the quantum potential does not "push" on objects, but informs their motion, structuring reality through its own form. Since it does not exert a force, the quantum potential is not directly detectable by physical means; one can only gauge its presence indirectly. Second, the quantum potential is instantly sensitive to all changes in the experimental arrangements, even very distant ones. The second feature is called *nonlocality,* and is a central feature of all present discussions of quantum theory, whether by Bohm or the quantum realists. We will encounter the concept again, and treat it more fully then. The Bohm and Hiley theory offers us the attractive possibility of imagining particle motion as we always have, but it comes at a cost. The cost is the introduction of a kind of "ghost" field, or immaterial ether— the nonlocal, quantum potential.

One may understand the photon to be a nonclassical, quantum object and so give up all meaningful statements about its history. Or with Bohm and Hiley one may populate space once again with a new quantum-mechanical, nonlocal ether. One view locates the quantum character of reality in a hidden environment (Bohm/Hiley), while the other integrates it into the photon itself. The former is a kind of dualism, the latter a monism. Yet both views insist that our world is not what it was once thought to be. One view gives up history in favor of a "quantum reality,"[7] the other advances a new "implicate order," as Bohm calls it, of which our reality is but a partial projection.[8]

Quantum phenomena show the world, at least on the atomic scale, to be utterly unlike the living-room world of sofas and armchairs. Its structure is other, its order unfamiliar; yet for all that, it is no less real. All conceptual schemes, no matter how they differ one from the other, must accommodate the data of archetypal quantum phenomena. To my mind, therefore, it is of far less importance which one theory we take to be true than to *see* what they all have in common. Each points from a different direction to a common core.

Walking around Michelangelo's *David,* we see it from different sides. Reading about Michelangelo's life, studying his contemporaries and his other sculpture, we enlarge our orbit around the *David*. Rather than fixate on one "view," should we not (like Einstein) hold on to our self while learning to see through the eyes of others? Similarly with regard to quantum phenomena. We should move among every equivalent theory, each interpretation, learning to see through its eyes. Each will highlight a different blemish of the old order and offer an individual remedy that reveals as much about its author as about the quantum. Yet for having circled the phenomenon in this way, something altogether wonderful happens. We change. We see *David* or the quantum phenomenon anew. Theories become *aids to reflection,* as Coleridge would have called them, not the codification of truth. Too quickly do theories become idols and so stand in the way of insight instead of promoting it. Moving among competing views, one is freed from the tyranny of monocular vision, and like the Indian god Varupa, encompasses the world with a thousand eyes.

What, then, are the essential insights into light offered us by quantum phenomena? We have begun, but to answer this question fully, we need to move farther around the statue. Already the outline is becoming clear. Notions such as history, place, and identity will need reworking, but in order to make these clearer we need to study other quantum phenomena. So far we have examined only single-photon, quantum effects. It is now time to go further to two-photon and many-particle phenomena, to enlarge our orbit as we continue to fashion new capacities for insight.

Entangled Light

The aim remains: to understand the world.
John Bell

Once again we will follow Einstein's lead. Deeply troubled by the implications of quantum mechanics, Einstein sought a way to demonstrate its incompleteness convincingly. In his view, quantum mechanics was only a partial picture of a far more subtle and complex reality, much of which remained hidden. If this were the case, then the peculiarities of quantum theory would be no surprise to anyone. Quantum theory never predicts individual events, but only the probabilities of them occurring. But incomplete knowledge always shows up as uncertainty in predictions, whether at Las Vegas gaming tables or weather predictions. Ignorance leads to chance. Yet is the converse also true—do uncertainties in experiment results always point up our ignorance? In the nineteenth century, a confident physics would have answered yes, uncertain results always reveal partial knowledge. Einstein insisted that rationality requires an unambiguous theory, and quantum mechanics is ambiguous and therefore not rational. It is not a proper theory at all.

To make his point, in 1935 Einstein, together with B. Podolsky and N. Rosen, once again resorted to a thought experiment, a brilliant one that has proved of unparalleled significance for the foundations of quantum mechanics, especially when, in 1964, its consequences were elucidated by a modest but brave young physicist from Ireland, John Bell.

—

PHYSICISTS WOULD AGREE that no single person has done more important work in the foundations of quantum mechanics than John Bell. His unique stature was due, I think, to his rare combination of deep modesty, profound courage, and unmistakable brilliance. Shortly before his untimely death in 1990, I had the privilege of organizing a workshop on the foundations of quantum theory together with my Amherst College colleague George Greenstein, but it was Bell's willingness to attend that brought people together. For a solid week he acted as an insightful and radically honest critic of quantum theory during our intensive conversations on its foundations. It proved a wonderful gathering, filled with humor and music, as well as helpful treatments of the major unresolved questions in the field.

As Bell emphasized again and again, for all its power, and few knew it better than he, orthodox quantum mechanics is simply not good enough. It will be superseded. Bell once wrote that the fate of quantum mechanics is apparent simply by examining its own internal structure. Look carefully and you will see its harsh future: "It carries in itself the seeds of its own destruction."[9] Einstein and Bell possessed the uncanny ability to see clearly the problematic malaise of quantum theory, and they both held tenaciously to the aspiration that a better and truer understanding of nature was possible. On the basis of Bell's theorems, some of the deepest issues in quantum mechanics became open to experimental examination for the first time. Those investigations are now reasonably complete, and their significance for our contemporary reimagination of light and substance is huge.

—

IN THE YEARS since 1975 when Bell proved his famous theorems, physicists have successfully performed Einstein's 1935 thought experiment, since dubbed the EPR experiment after its authors Einstein, Podolsky, and Rosen. It is an archetypal experiment which, if understood fully à la Bell, carries us far into the new order of things. It asks after reality, seeking for the "be-ables" of a theory, as Bell termed them. What are the real properties that inhere in things? Especially for light, the answer leads to a far more subtle and, I think, beautiful understanding than ever before. We begin very simply.

Objects of this world bear attributes: my pen has color, shape, mass, and so on. Every object must have some well-defined attributes by which we know it. Thus, my pen cannot be red and green at the same time, without me suspecting my sanity. Without sensible attributes, an object loses its identity; in a sense, it disappears. Einstein asked, what are the real attributes of quantum objects? By these we shall know them, or we will not know them at all.

Since quantum objects are delicate things, one needs a means of testing for an attribute that is noninvasive, to use medical terminology. After all, if one is not careful, the object may be disturbed so violently that it changes in one's hands. Then, as if wearing rose-colored glasses, I will see not the object's color, but the color of my glasses. In order to avoid this problem, Einstein suggested the following. Let the objects of study be produced in pairs, like identical twins born together and indistinguishable except for one attribute. Staying with the analogy of the twins, let the one distinctive attribute be a birthmark, under the left arm, say. By this mark alone can we tell who is the prince and who the pauper.

The twins rise each morning, bid each other farewell, and head in opposite directions to laboratories at opposite ends of town. One twin (we don't know which) reaches his destination slightly before the other. A scientist, uncertain as to who he has before him, checks for the hidden birthmark. In that instant, his uncertainty disappears as he recognizes the lad. Not only does he know with certainty who he has before him, but can also predict with absolute confidence, without observing the distant twin, who will show up in his colleague's lab a few minutes later. In a sense, observation on one twin is sufficient to make true statements about both. The scientists keep track of their results, X's for pauper (birthmark), O's for prince (no birthmark).

SCIENTIST A		SCIENTIST B	
Feb. 1	X	Feb. 1	O
Feb. 2	O	Feb. 2	X
Feb. 3	X	Feb. 3	O
Feb. 4	X	Feb. 4	O
Feb. 5	O	Feb. 5	X
. . .		. . .	

Notice the perfect correlation; whenever one is X, the other is O. This is precisely how we expect the world to behave. Each twin wends his way to one laboratory; one has a birthmark and the other does not, one is the pauper while the other is the prince. All is well with the world.

Einstein proposed a quantum version of this experiment, the EPR experiment. Forty years later, Bell proved that the predictions of quantum mechanics and those of any, alternative, "local, realistic" theory would differ in specific measurable ways. The important words are "local" and "realistic." What is a local, realistic theory? They are just the theories that give explanations about which we would feel good. They are commonsense theories. In addition, Bell's theorem is *very* general. It does not advance one competitor to quantum mechanics, but all commonsense competitors in one fell swoop! The stakes are very high. On the basis of Bell's theorem, one good experiment can eliminate every commonsense competitor to quantum mechanics.

The experiments have been done, most beautifully by Aspect and his collaborators in France. They undeniably show the quantum-mechanical predictions to be correct, not those of any Einsteinian, local, realistic, commonsense theories. Experiments necessitate, therefore, that something in Einstein's view of rationality must be given up. The question remains, what? The EPR/Bell archetypal experiment urges on us the possibility of a more flexible form for rationality than that of traditional science. We need not give up rationality, but rather must broaden its meaning. Our conceptions of rationality were unreasonably limited by our bias toward a mechanical universe. After all, shall we declare the mathematics of quantum theory "irrational"? Mathematics can perform with complete calm the most extraordinary things. Let us turn to Einstein's quantum thought experiment, and see in what ways rationality needs enhancement.

Briefly stated: In laboratory EPR experiments, two photons take the place of the twins. They are produced simultaneously and also carry an attribute or "birthmark" called polarization. Moreover, they are produced in a special way such that if one is X-polarized the other is O-polarized, or vice versa. If we do not check their polarization, then, as for the twins, the photons are indistinguishable. We allow the photons to fly off in opposite directions to distant detectors that are able to measure polarization. One arrives and its polarization is checked. If it proves to be X-polarized, then the scientist there *knows* the other is O-polarized. Repeat the experiment many times. As before, the data from the two detectors are found to correlate exactly; for every X there is a corresponding O, for every O an X. Einstein declared any attribute for which such sure predictions are possible to be a *real* attribute of light; that is, it existed even before the measurement was made. After all, how can distant actions on one twin or photon immediately affect the properties of the second? Effects act locally, not at a distance. In order to affect a distant object, a signal must travel the entire distance, and that takes time. Again, this is common sense, an obvious feature of any good theory, according to Einstein. He termed it "locality." Commonsense theories are realistic and local.

The correlations above are much like those for the twins. Quite naturally we would initially interpret all this to mean that each photon is *really* polarized from the outset, just as each twin really is marked or not at birth. All this is completely reasonable; it is exactly the way a rational world should be arranged. Unfortunately, for correlated quantum particles, matters are otherwise! Quantum attributes are not so neatly separable, our actions are not so nicely localized. Bell and the recent EPR experiments drive this point home unforgivingly. In general, light correlations do *not* behave like twin correlations under all circumstances. Bell's theorem pointed out those special situations for which quantum theory and local, realistic theories differ, and subsequent measurements have come down clearly in favor of quantum predictions. Common sense is thus violated in an unavoidable way, but what is the significance of the violation?

Einstein's view was one in which objects possess real, enduring attributes such as color, polarization, and a path of travel. EPR experiments, especially the most recent ones by Aspect and collaborators in France,[10] measure the attribute of photon polarizations for various detector orientations. Their results, when coupled with Bell's theorem, convincingly demonstrate that there is *no local, realistic* way of understanding polarization correlations! The experiments and logic that lead to this conclusion are compelling. They demand of us an extremely deep reformulation of what we take light to be. To make clear the grand implications of these pivotal experiments, we can examine their implications for theoretical interpretation. Again two prominent tacks are taken. Both are important aids to reflections as we ponder the archetypal quantum phenomena of light.

—

THE FIRST AND most common response is that paired photons cannot be understood as individually polarized (i.e., marked) from birth at all! In this understanding of quantum mechanics, the attribute of polarization is no longer granted to each photon of the pair separately, but is somehow only a shared or holistic property of a new kind of object. Here is an instance in which the whole is rigorously *not* the simple sum of its parts. I will call this view "quantum realism." According to it, one cannot conceive meaningfully of two separate photons traveling through space each with its own unique and enduring polarization. Quantum reality is different; separability is lost, and the relationship of whole to part is not that given by Kant to the inorganic world. Once two objects have interacted quantum-mechanically, they join to become a new single, "entangled" entity, as Schrödinger called it. The new entangled state of light does not travel in the usual sense, but evolves in a more holistic way, retaining its ambiguous nature throughout.

At this point an enormous problem arises, one seen early on in the development of quantum mechanics. If quantum reality is entangled in surprising and subtle ways, how is it that we see sense reality as separate and disentangled? This is the "measurement problem." Quantum reality is whole, all of a piece *until* a measurement is made. Then somehow what was one becomes two; what was an entangled whole becomes disentangled parts, and does so instantly. In the case of paired photons in the EPR experiments, an exact polarization is found at one detector, and in addition, a distant paired photon simultaneously shows up with an exactly correlated polarization! The ambiguous, entangled state is somehow reduced, and the reduction can take place over arbitrarily large distances, instantly. Clearly, Einstein's condition of locality is violated. How does orthodox quantum theory account for this nonlocal reduction of entangled light, that is, for the simple fact of measurement? It cannot. Many have been the proposals, past and present, but none has yet succeeded in addressing the measurement problem convincingly. If quantum reality *is*, then the passage from it to sense reality is miraculous. At this point several physicists and philosophers, including the Nobel laureate Eugene Wigner, introduce the active agency of the mind. The miracle of reduction, they say, takes place by the action of mind in the moment of cognition. Only when one fully includes its role, they argue, can one account for knowledge.

Turning aside from the measurement problem, at least for the moment, we can search for a surer footing with regard to the nature of light. As proved by experiments with photons, the attribute of polarization does not have a unique, enduring "local reality" in the Einstein sense. Perhaps the EPR identity problem is only associated with the attribute of polarization. Light is specified by characteristics other than polarization, which (we hope) may not suffer the same ambiguity. Perhaps these other attributes will be firmer, sharper, and less illusive. For massive particles, for example, the attribute of electric charge is unambiguous (a so-called super-selection rule requires charge never to appear in an entangled state). But the photon has no mass, or charge. In fact only four attributes are used to define light formally; they are polarization, wavelength, direction, and intensity. We have already seen that the history of a photon's path is ambiguous, so direction is no better than polarization. Experiments show that wavelength (i.e., color) and intensity also suffer exactly the same fate as polarization.[11] All can become entangled, one color with another, and so on. The conclusion: There is no truly unambiguous attribute of light! Today's quantum-optics experiments challenge the root conceptions we have about the separable atomistic structure of the world. Moreover, much the same can be said to hold for material particles. For example, even the mass of objects can be put into the same kind of entangled superposition state that light evinces.

How bad is this? Very bad. If true, and *it is true,* then it implies there are parts of reality (assuming there is a reality) where attributes do not map onto things simply. It would be as if you found an object that was no particular color, shape, size, mass, etc. It was nothing in particular, but still remained a very specific thing. What are the attributes of light? It seems a simple question, and quantum theory gives what is at first a simple answer: polarization, wavelength, direction, and intensity. More careful thought, however, shows that quantum reality handles the attributes of light far differently than does sense reality. Sense objects must possess sharply defined attributes. Light, quantum mechanically considered, need not. Its attributes are more holistic; in general they exist in inseparable or entangled combinations, at least until the moment of measurement, whatever that is.

Ever since Galileo, Descartes, and Newton, science has sought for the "primary qualities" of things, that is for the unambiguous and irreducible attributes of reality. The senses provide only secondary experiences, but behind them, they argued, lay the primary qualities of extension, or mass, or solidity, etc. What are the primary qualities of light that vouchsafe its unambiguous existence? The extraordinary response given by quantum realism is that there are none. Light, as an enduring, well-defined, local entity vanishes. In its place a subtle, entangled object evolves, holding all four of its quantum qualities suspended within itself, until the fatal act of measurement.

—

WHAT ARE THE responses of physicists to these metaphysical implications? Foremost is what Bell has called FAPP, "for all practical purposes." We have a practical way to proceed, they say, and would do better to give up all hope of an objective picture of reality, a suggestion strongly advanced by Bohr. The procedures laid out by quantum theory are adequate "for all practical purposes." The troubling areas are not large and mostly philosophical, it is said, and can often be made nearly to vanish (although they never disappear entirely); so FAPP physics is good enough. Schrödinger protested in the strongest terms.

> A widely accepted school of thought [Bohr's] maintains that an objective picture of reality—in any traditional meaning of that term—cannot exist at all. Only the optimists among us (and I consider myself one of them) look upon this view as a philosophical extravagance born of despair in the face of a grave crisis. We hope that the fluctuations of concepts and opinions only indicate a violent process of transformation which in the end will lead to something better than the mess of formulas that today surrounds our subject.[12]

The grave crisis will not go away by ignoring it. Here are the seeds of self-destruction referred to by John Bell as latent in quantum theory itself. Besides, there are many interesting avenues to explore, exciting implications to play out, even if they are uncomfortable for traditional commonsense realism. The world may very well prove to be rational in the end, but its structures are far more varied than those immediately revealed to our senses. We should learn from Bell, Einstein, and so many others to be awake to what confronts us, to face it clearly and carefully, and to be ready to transform boldly our most habitual ways of seeing.

—

LIGHT FALLING ON the eye provokes sight. Until that moment, light lives in a universe of its own, all its commonsense properties caught up into itself. As we saw in the last chapter, even space and time lose their meaning when one imagines traveling at the speed of light. Intervals of time and distance disappear. As light touches every atom it enfolds atomic history into its own, entangling itself ever more widely. Ironically, as it does so the drama of the EPR results in one way declines. Quantum theory usually predicts the largest divergences from common sense for the simplest entanglements of light. Quantum effects are generally diluted as entanglement grows, disappearing for-all-practical-purposes (FAPP). Yet even this is not always true, and when it is not, the drama only increases.

Every single-photon interference experiment, every two-photon EPR experiment, and many I have not touched on have also been performed with matter, that is with electrons, or other atomic particles. For example, neutrons, the uncharged constituent of the nucleus, have been used recently in magnificent demonstrations of wave-particle duality in just the way I have used photons.[13] Most recently, several experimental research groups have even used atoms, many times more massive than neutrons, and placed them into superposition states that show the same interference effects as massless photons.[14]

In addition, a few many-particle experiments have now been performed that convincingly show that the stunning effects due to entangled states do not always diminish with increased numbers of particles. The experiments at IBM's Watson Research Center by Webb, Tesche, and Washburn have woven billions of electrons into collective, entangled states to produce impressive quantum-mechanical effects.[15] High-temperature superconductors also promise to bring such collective quantum states ever closer to everyday life. The remoteness of quantum paradoxes may not be long-lived. Then what will we make of our world?

Quantum realism holds that the world really is the way orthodox quantum mechanics says it is. Properties of photons such as path and polarization do not exist properly before measurement. Ambiguity disappears in that instant by mechanisms not yet known to us. Many are the problems with such a view, but what are the alternatives? There are several, but to my mind the most enduring has been, and likely will be, that of David Bohm's mentioned before.

For every experiment performed to date, Bohm has a consistent theoretical account. In every case, a "real" photon travels a real path, with a real polarization and wavelength. The entanglement of quantum realism is not, in his view, an entanglement of quantum particles, but of his new quantum potential. The new physics of quantum mechanics is carried by it, including the mysteries of nonlocality. The electron, neutron, or photon is guided by this ghost field. If one grants Bohm the existence of his quantum potential, then he can explain every experiment with the additional benefit that even measurement is accounted for! The world of sense reality, of particles and fields, is very like it has always been, but it is understood by Bohm to be a projection of a far more subtle "implicate order"; the holism of EPR and entanglement are essential aspects of its nature.

Bohm's is an attractive view, but requires adherents to accept the profoundly new implications it entails. Among these are the quantum potential, and the existence of a source of energy within all elementary particles. Bohm argues that the quantum potential does not "push" or "pull" objects the way other forces do (for example, gravity), but rather the quantum potential "informs" their motion as the movements of a remote-controlled model airplane can be directed from the ground. Very little energy is needed to send the signal to the airplane, but the consequences of its reception can be dramatic. The behavior of quanta are like those of the plane, which are guided by the quantum potential.

At this point, several problems arise. No one has ever found direct evidence of the quantum potential. The tiny electromagnetic signals that guide model airplanes and space probes can be detected with sensitive equipment, as they are on the crafts themselves. Despite experimental searches, why have we never found the "ghost field" of the quantum potential? If it is real, then it must transmit at least a small amount of energy to the receiving particle. That energy should be detectable, but has never been seen. In addition, the model airplane has an engine and servomotors internal to itself. These are directed by the information carried by the radio signal. Bohm's view implies, as he himself says, that every elementary particle possesses an internal "engine." Particles appear inert at one level, but at the level of the quantum they are "self-movers." Yet all experimental evidence to date, even at the highest energies and smallest dimensions, has turned up no evidence of structure in the electron or of "engines" within elementary particles. It is nearly impossible for most physicists to imagine an engine within the photon or electron, or that the ghostly quantum potential can be so all-pervasive and yet so elusive. To them, FAPP and quantum realism, even with the measurement problem, are easier views to adopt. Still, to my mind, Bohm offers a serious alternative, even if few, far too few, have joined him in his research into implicate-order physics.[16]

—

IN EITHER VIEW, that of Bohm or quantum realism, a monumental shift has taken place in our conception of things. It goes by the humble name of "nonlocality," but within it is concealed a revolution in thinking. No new predictions will arise from it, no new high-tech gadgets, but if taken seriously, it urges us to do what Goethe, Thoreau, Whitman, and a thousand others have petitioned for a century: to conceive of things whole. As Francis Thompson phrased it,

> *All things . . . linked are,*
> *That thou canst not stir a flower,*
> *Without troubling of a star.*[17]

If one day we awoke and truly saw the world this way, the ramifications would shake our psyche to its foundations. Assuming we remained sane, the relationship of I to thou, of individual to planet, of my actions to yours, would be revolutionized. Edward Lorenz's theory of nonlinear systems delivered the "butterfly effect." The flight of a butterfly in Rio de Janeiro can, in fact, change the weather of Japan. If chaos dynamics has shown us the extreme delicacy of our world, quantum physics reveals its deep intimacy.

Every imagination of light has been held within a larger cultural imagination of man and world. Are we now at the watershed to a new one, and could it possibly support a true ecology of human, animal, plant, and mineral communities? In recent decades, light has taken on a new and subtle figure; one can only hope it is but a symptom of a larger evolutionary change in the structure of our imagination that supports an ecological consciousness.

—

WHEN I TRY to imagine light without a particular color, direction of propagation, etc., I understand the struggles of medieval theologians or artists as they sought to picture God, and appreciate their choice of light as an attribute of the divine. God's existence was not at issue, but his features were. Anything definite one said about them, any image that depicted Him, was of necessity partial and potentially misleading. Even such a simple thing as the location of God, where God is, was full of danger. Amazingly enough, light, too, suffers from a special difficulty with regard to location. We have seen its other attributes entwined in subtle ways, but what of the simple sense of place? As the next section shows, the very concept of place loses meaning for light as for no other object.

The Place of Light

*Hast thou perceived the breadth of the Earth? declare if
thou knowst it all. Where is the way where light dwelleth?
and as for darkness, where is the place thereof?*

Book of Job

In his story of world creation, before anything else could happen, Hesiod had first to bring forth a brooding space filled with possibility, called Chaos. In Greek, its root meaning was "gap." Here, in the place provided by Chaos, rose the "wide-bosomed Earth," as Hesiod called it, "a sure, eternal dwelling-place for all the deathless gods who rule Olympus's snowy peaks."[18] The place of earth was created before the earth itself.

Everything must have location, a place where it is. Aristotle declared that "A natural scientist must inform himself not only on the infinite, but also on place."[19] What then is the place of light? One would suppose that with the photon concept of light, a straightforward response would be forthcoming; but quantum theory and experiment again conspire to make the place of light completely elusive.

—

IN AN IMPORTANT paper of 1949, Eugene Wigner and T. D. Newton, then at Princeton, looked for the place of elementary quantum particles: electrons, protons, mesons, and photons.[20] They were interested in "localized states," that is to say, they sought a clear way to define position for particles within the formalism of quantum mechanics. Everything began well enough. Elementary particles that possess mass, like the electron and neutron, presented no special problems. However, when they turned their attention to light, a sudden hitch appeared. They commented that light was different; they could find no mathematical object within quantum theory that properly corresponded to the concept of place or position as we know it. That observation, their trouble in finding the place of light, persists to this day.

In their analysis of the laser, Marlan Scully, Murray Sargent, and Willis Lamb carefully describe how light is reflected back and forth between the two mirrors of a laser cavity. If light could be conceived of as corpuscular, we would naturally picture to ourselves a kind of tennis match with balls of light bouncing back and forth in the laser. This is *not* the case! In their words: "Photons are not localized at any particular position and time within the cavity like fuzzy balls; rather, they are spread out over the entire cavity. In fact no satisfactory quantum theory of photons as particles has ever been given."[21]

Many times over the last sixty years, the quantum theory of light has been searched for a way to give a position to light, and it has consistently denied what it otherwise provides so readily for massive particles. Why is light so recalcitrant about declaring its position? The answer seems connected to the transverse nature of the electromagnetic field. Remember Fresnel's discovery that polarization could only be explained by considering light to be a transverse wave. This innocent observation, when carried over into the quantum theory, makes the definition of position impossible.

It does not mean that there is no commonsense meaning to the place of light at all, but it is always limited in a critical way. There have been several experimental tours de force in recent years that attempt to locate light's position in one way or another. One of the most dramatic and easiest to understand is what is called "light-in-flight" or LIF holography. First demonstrated in 1978 by Nils Abramson in Sweden, the technique has since been developed so far that it can produce "stop-action" images of traveling light pulses no thicker than the breadth of a hair.[22] Yet what one sees here is a wave front, not a particle of light, that is to say, only one spacial coordinate is well defined.

By contrast, a massive particle has sharp values for all three spacial coordinates. Not so light.

Interesting recent developments have connected an EPR-type experiment to the question of location, not only for light, but for all elementary quantum systems. In 1989 J. D. Franson of Johns Hopkins University suggested a variation of the EPR experiment in which the ambiguous attribute is not photon polarization, but its time of emission. In terms of our twins analogy, the ambiguity would now be in their time of departure. It is important to realize that one is not simply ignorant of the departure time, but far more radically, that departure time *does not exist* as an unambiguous feature of the twins. The ambiguity in the time of photon or particle creation results in a parallel ambiguity in its position. Franson's argument applies to massive particles, such as electrons and neutrons, as well as to light, as long as they are in an entangled state. But in addition to the fundamental ambiguity entailed by his considerations for all particles, light has the added problem that even for disentangled states, the concept of position is invalid.

By now it should be evident that light possesses a nature unique to itself. Every natural assumption we make about it, assumptions common to us from daily life, leads to errors. Entering light, we cross into another domain, and must learn to leave behind what we hold dear from the past, and cleave only to the archetypal phenomena of light at every level, down to the quantum. Particles, waves, location . . . all should be left, like soiled sandals, at the threshold of the temple. The light within is of a different order than the objects without; it inspires us to subtle reflections not common to the marketplace. We stand like Brunelleschi, in the portal between sanctuary and piazza. He looked out, intent on the geometry of sight; we are turned in, absorbed in the morphology of light. From the union of our open imaginations with the firm facts of light will rise the offspring of insight into it.

—

THE CONTEMPORARY ARTIST James Turrell has said, "Light is not so much something that reveals, as it is itself the revelation." Holding to light, not to the objects it illumines, is the point. We need time to work with it, live with it, think about it, and see into it.

Turrell has rightly been called a light smith, forging his sculptural art out of pure light. His installations are architectural spaces made bare of all but the essentials so that light itself can become the object. Everything leads to light. Turrell speaks of his working with light as an attempt "to create an experience of wordless thought," an engagement with immaterial reality. "I have an interest in the invisible light, the light perceptible only in the mind. A light which seems to be undimmed by the entering of the senses. I want to address the light that we see in dreams. . . ."[23] I have a similar aspiration, that each of us become a light smith.

Turrell senses the mystery of light's insubstantial power and fashions artistic forms with that energy. Light, he says, "has a quality seemingly intangible, yet it is physically felt. Often people reach out to try to touch it. My works are about light in the sense that light is present and there; the work is made of light. It's not about light or a record of it, but it is light. Light is not so much something that reveals, as it is itself the revelation." In this and preceding chapters, I have made an analogous attempt, drawing into view the selfless, ever-present, but elusive actions of light.

I have written of light's life, it remains for us to form a picture of its character. We cannot do so abstractly but only by drawing together the many sides it has shown us, by moving through one aspect to the next, always asking, what is it that can show itself in so many ways? We have watched the transformation of light from a divine presence to a material object, we have seen the body of light become ever more subtle, immaterial, and para-doxical. Every attribute of light can and does become entwined, entangled, and nonlocal. Color, place, polarization, intensity . . . everything by which light is classically named loses its common significance; a new name is needed. Circling through the life of light, we can hold its essence only gently, as we might hold a fledgling bird ready to fly. Fitted for air and space, it touches the earth awkwardly. Following it in flight, running with light, we seem caught up into its mysterious nature, at once here and there, a connective tissue weaving together all of existence, a whole whose parts are wholes themselves, a thing for which time and space disappear. I cannot describe it, my imagination can only just touch its hem, but I do know that at its core there seems to live an original or "first light" within which wisdom dwells, a wisdom warmed by love and activated by life. Around it a many-roomed mansion has risen, and our wanderings there have not exhausted its riches.

—

As WE LEAVE light's expansive dominions, the heavens dim and darkness quietly falls. Within that darkness there is a silent murmur, a still voice that whispers of yet another and unsuspected part to light, for even utter darkness shimmers with its force.

Dark Light

Yet mystery and manifestations arise from the same source. This source is called darkness. . . . Darkness within darkness, the gateway to all understanding.

Lao-tzu

On May 18, 1885, the French poet Victor Hugo at age eighty-three had a stroke. Four days later, during his death struggles, he, like Goethe, spoke of light, saying, "Here is the battle of day against night." Hugo's last words continued what in life he had always done: searched the darkest recesses of human nature for its brightest treasures. As he died he whispered, "I see black light."

Is there a light in darkness? Is the night empty, void, and dead, or does it, too, offer more than appearances suggest? If we trusted the poets enough to ask, their reply would be unanimous. Whether it be Novalis, Goethe, Hugo, or Nerval, the night has always invited the poet to descend into its fathomless seas. Beneath dark waves every stoke can strike a thousand sparks, like the bioluminescent glow of a sailboat's wake. Dark seas shine in poetic imagination, and in journeying to the sublime, one enters through the doorway of darkness. The French poet Gerard de Nerval wrote of it this way: "In seeking the eye of God, I have seen a vast sphere black and fathomless, whence the night that inhabits it shines forth on the world and continually deepens."[24] The night shines forth with its own dark light, entreating the bard to read by its radiance, the eye of God.

Distrusting the poet's perceptions, we turn to the physicist for levelheaded counsel and are perhaps surprised to learn that they, too, speak of light in utter darkness. I am gratified, because I would have all physicists be poets. Yet what is the physicist's dark light, and how do they sense it?

—

DURING A 1948 walk with Niels Bohr, the Dutch physicist H.B.G. Casimir told Bohr of a difficult calculation he had just performed. It had to do with the possibility of a force of attraction existing between two uncharged metal plates. The calculation had been long and arduous, but surprisingly, the result was wonderfully simple in its final form. The plates should attract each other inversely as the fourth power of their separation. How could something so simple arise from such a complex analysis? Casimir was suspicious. Bohr agreed and pointed Casimir in a totally unexpected direction. Casimir's calculation had examined the detailed mechanisms by which atomic motions in the metal plates might induce unexpected electromagnetic fields and thus attract a nearby plate.[25] He had used the still relatively new quantum theory of Schrödinger and the relativity of Einstein, but Bohr urged Casimir to turn his attention away from the plates and to the empty space around them. It would have been senseless advice twenty-five years earlier, but in the intervening decades a full quantum theory of light (quantum electrodynamics) had been developed, and one of its features was a new understanding of the vacuum, of emptiness.

Where before the vacuum had been understood as pure emptiness—no matter, no light, no heat—now there was a residual hidden energy. Take away everything, cool to absolute zero in temperature, and still the vacuum remains, and it is shimmering with a special kind of light. Called the "zero-point energy of the vacuum," it seems an essential part of quantum-field theory. Bohr suggested that Casimir look to it, to the vacuum, for the force between two metal plates. Casimir followed Bohr's prescient advice and in a two-page derivation, he arrived by an elegant path at what had caused him so much trouble before.[26] Since Casimir's calculation, experiments have shown the force to occur with exactly the form he predicted.

Without launching into the details of quantum electrodynamics (or QED), I can still draw attention to a few features of the calculation that are suggestive of a new way of understanding emptiness. According to QED, after one has removed all matter and all light from space, an infinite energy still remains. Since there is no way to extract this energy out of the vacuum, theorists dismissed it as a curious artifact of the theory, of little real significance, at least until Casimir made his calculation. By inserting two parallel conducting plates into the vacuum, Casimir showed that the structure of the vacuum could be changed. Move one of the plates and it changes again.

One can calculate the zero-point energy of the volume between the plates for both plate separations. It is infinity both times! However, Casimir realized that by subtracting the one infinity from the other, carefully, he could obtain a finite result. Infinity minus infinity can, in some cases, give a reasonable, finite result. This calculation proved significant not only for the specific predictions it made, but also because it showed physicists how to eliminate the infinities popping up in their QED calculations.[27]

For our purposes, Casimir's problem suggests another way of understanding darkness. Darkness may be a far richer, more subtly structured fullness than we have thought until now. Recent theory and experiments only encourage these musings further. Even in the deepest shadow, we can seek, and perhaps find, a hidden light.

—

VICTOR HUGO SAW darkness as a twofold being, divided between the fallen Islamic angel and tempter of man, Iblis, and the heavenly human being he called Christ. Both belonged to the night: to one belonged the emptiness of despair, to the other the true comfort of compassion. The key to each is within us. At seventy-five, Hugo wrote of darkness with the key in his hand.[28]

> *O darkness, the sky is a gloomy precinct*
> *Whose door you close, and whose key the soul owns;*
> *And night divides itself in half, being diabolical and holy,*
> *Between Iblis, the black angel, and Christ, the starry Human Being.*

It is empty, and yet within its gloomy precinct lives the holy, starry Human Being.

—

WE HAVE TRAVERSED an enormous span in our natural history of light, both scientific and sacred. Like a jeweler studying a faceted diamond whose fleeting colors change with the smallest turn, we have held light always before us, rotating it slowly before our eyes. It has taken on a thousand forms, changing in our hands like the Greek god Proteus. Yet for all that, it is one being still. Paraphrasing Herder, we can say, "That which is called light in creation is, in all its forms and in every being, one and the same spirit, a flame unique."[29]

Across from that flame unique, however, has always been a changing face. First the countenance was that of an ancient Egyptian, then a Greek, a Manichaean, a Cathar, a Scholastic bishop . . . a physicist and ours. Each culture brought its own question, each the wealth and limitations of its imagination, and the flame unique lit each according to its own nature. The history of light is not a steady convergence to truth, nor does it end in a relativism that makes inquiry meaningless. Insights into light belong to every culture; truth has many homes. The history of light mirrors the history of mind, and continues to offer fresh aspects to new awakenings of the soul.

Again and ever again, it is ourselves whom we study in studying light. Thus we uncover the evolution of that knowing consciousness we prize so highly, but whose life and growth we too often neglect. After we accept the history of its changing character, where does mind now tend, what are the possibilities for its conscious cultivation, and what will be the new fruits of tomorrow's awakenings to light?

CHAPTER 12

Seeing Light

*Our whole business in this life is to restore
to health the eye of the heart whereby God
may be seen.*[1]

Augustine

"The divine light penetrates the universe according to its dignity," wrote Dante in his *Paradiso*.[2] To the medieval imagination, as light penetrated the realms of stone and glass, it dignified town and village by calling into being the Gothic cathedral. Light and geometry were the intertwined themes of these sacred structures. For the first time, an architecture had been discovered that could free walls of their structural loads so they could become the jeweled membranes of a Heavenly City. Sacred geometry ordered the shape, but light, like the breath God breathed into Adam, gave the cathedral life, swelling it just as the Virgin's womb swelled with the Light of Mankind.

As St. Bernard conceived it, the passing of light into and through stained-glass windows was a literal reenactment of the conception and birth of the infant Jesus by the Virgin mother. In both instances the material kingdom was penetrated but not violated. "As a pure ray enters a glass window and emerges unspoiled, but has acquired the color of the glass . . . the Son of God, who entered the most chaste womb of the Virgin, emerged pure, but took on the color of the Virgin, that is, the nature of a man and a comeliness of human form, and he clothed himself in it."[3]

Two worlds, the physical and the spiritual, were one within the sanctuary of the cathedral; and light played a leading role in that unitary imagination of God in man's domain. The great art historian of the Gothic, Otto von Simson, wrote,

> With its sublime theology of light it [the Mass] must have conveyed to those who listened, a vision of the eucharistic sacrament as divine light transfigured through the darkness of matter. In the physical light that illuminated the sanctuary, that mystical reality seemed to become palpable to the senses. The distinction between physical nature and theological significance was bridged by the notion of corporeal light as an "analogy" to the divine light.[4]

There are levels to light, and more than one may be active at any one moment. Some, from the past, may be alien to us, and their union with others that are more familiar may be disconcerting. Yet, while we may be deeply confused, the culture we study seems oblivious to our dilemma. One feature of scientific progress has been the segregation of levels once integrated, for example, the isolation of values from scientific knowledge, the photon from the Incarnation. This separation has had its costs as well as its benefits.

—

TRADITIONAL CULTURES, as well as the early history of our own, were, to use the language of the Cambridge anthropologist Ernest Gellner, "multi-stranded."[5] When a member of the Nilotic tribe of the Nuer looks at a cucumber and, in complete seriousness, identifies it as a bull, he entangles himself in no logical absurdity because he lives in a multistranded mental world. The strand of cucumber as a vegetable food is woven into, but not confused with, the strand of cucumber as totem. The history of the West is a history of the growing segregation of the strands of consciousness, the separation of the moral and spiritual from the sensual and physical, the loss of the unity felt by the Nilotic clansman.

The felt unity of the spiritual and physical that still pervaded the consciousness of the thirteenth century reflects itself in early understandings of history. The outer, historical events of the Old and New Testaments were simultaneously a revelation of the spirit. Entering a Gothic cathedral, one passed through a portal whose intricate stone statuary rendered visible the history of God's actions on earth. In Chartres, the window depicting the genealogy and birth of Jesus, the beginnings of things, stands above the entryway. Moving down the nave aisle, one passes between images drawn from the history of the Jews and the miracles of Christ's life. Where transept crosses nave, the sacrifice of Christ crosses the linear axis of history through which one has just passed. The altar railing delimits the spaces of present and past, of the Shepherd and sheep, the priest and his flock. Often, the iconography of the apse rising above and behind the altar is that of the Last Judgment and new Jerusalem, our eschatological future. The journey is through time, conceived of as at once secular and sacred. The dismembering of this unity is a relatively modern event.

Every early culture has understood its genesis and history as one woven of threads both divine and mundane to form a multidimensional mythology of creation, destruction, and migration. The organization of the drama is temporal. In Hindu cosmologies, world time is ordered into "yugas," the Maya have their "katuns," the Aztecs their many ages and five "sun" periods. Among the Greeks, Hesiod denominated the ages of man as Golden, Silver, Bronze, Heroic, and Iron, the last of which is our own. Even our word "world" stems from the Old Germanic compound *wer-aldh* meaning "the life or age of man."

To each age was associated not only external events, but a moral order as well. The Golden Age, writes Hesiod, was a time when men "lived like gods, carefree in their hearts, shielded from pain and misery. . . . They knew no constraint and lived in peace and abundance as lords of their lands."[6] After this blessed age, a second childish and unnatural race arose, followed by a third brazen race of mortals, who were a harsh, violent people, full of destruction. Then came the "divine race of heroes," among whose number were Achilles and Odysseus. The fifth race of men, our own, is a tragic mingling of justice and injustice, a time of disorder, when all holy alliances between families and friends are violated, and the hallowed patterns of life are lost completely. Hesiod's was a mythopoeic, multistranded history of human nature, not a chronology of purely mundane events.

The rich, psychospiritual story of our origins has only recently given way to another imagination. Since the rise of science in the sixteenth and seventeenth centuries, the physical origins of the world have gradually disentangled themselves from its spiritual origins. Divine cosmogony found a rival in the upstart physical cosmogony of astronomy and physics. With the appearance in 1859 of Darwin's *On the Origin of the Species by Means of Natural Selection*, the battle moved closer to home, from the realm of cold matter, planets, and stars, to plants, animals, and Homo sapiens.

In purging our history of the sacred, we have also unwittingly lost sight of our cognitive role in history. In our struggle for an accurate chronology of events and a description of influences, we have neglected the minds, or mentalities, of the figures who enacted the history. We have neglected the "soul of history." At best we give a chronology of ideas, an intellectual history, but we leave the history of their thinking (as opposed to their thoughts) untouched.

Implicit in life in every culture is a social and symbolic system, a tacit knowledge that, at least in part, is used to construct reality. Over two hundred and fifty years ago, the great

Italian philosopher Giovanni Battista Vico recognized the entanglement of mind and society when he wrote that "the world of civil society has certainly been made by men, and that its principles are therefore to be found with the modifications of our own human mind."[7]

Light has lived through all ages, yugas, and societies. Its transmutations exemplify a dramatic evolution in consciousness. What has been true for peoples is also true within the course of an individual life.

Psychogenesis

Then again, do you not see the year assuming four aspects,
in imitation of our own lifetime? For in early spring it is
tender and full of fresh life, just like a little child. . . .
                                                              Ovid

In both Utica, New York, and Washington, D.C., hangs an allegorical series of four paintings by the American artist Thomas Cole, in which the seasons of life from infancy to old age are imaged. In the early springtime of life, an infant appears in a small boat whose shape has been sculpted into the figure of Hours. At the helm stands a radiant angelic being who pilots the craft and her young charge out of a dark cavern into luxuriant growth bathed in a misty dawn light.

When, in the next panel, the infant has become a youth, the landscape opens up into a vast, exotic, and exciting prospect. The youth, now fully animated, yearning for the multitude of dreams outspread before him, takes the helm, while his unnoticed spirit guide gestures farewell from the bank. In the third canvas, *Manhood,* the boat is poised at the brink of a barren and dangerous cataract. The helm is broken, the sky is dark, the light is menacing, and the now-mature soul seems utterly lost. Only from the heights at the upper left of the painting does

The Voyage of Life: Childhood by Thomas Cole.

The Voyage of Life: Youth by Thomas Cole.

there stream a faint light of hope in which we can make out the delicate shape of his angelic companion. The final season of life, *Old Age,* is rendered as a vast and lifeless expanse of rock and water; "the stream of life has now reached the ocean to which all life is tending," as Cole himself characterized it.[8] The figure of Hours is broken and gone. Seated, the timeworn and bearded figure certainly recognizes that his journey through life is at an end. For the first time, he sees the spirit who has accompanied him throughout. It gestures toward a brilliant field of light, his luminous future.

In these four canvases, time has become space. As if at a family reunion, or perhaps in a traditional peasant village, all the ages of man are simultaneously present. By simply shifting one's gaze, decades can lapse, moving from infant to grand-mother, from worried father to youthful daughter, all with the turn of an eye. The full gamut of human experience is drawn into the lines of every face, and into the creases of every hand. These are the images of hope and care, of perennial aspirations and the harsh knowledge of mortality, set side by side.

Cole's cycle made a deep impression on the American public. Shortly after they were hung, a man of middling years was found viewing them alone. He had been looking a long time before remarking, with a melancholy air, "Sir, I am a stranger in this city, and in great trouble of mind. But the sight of these pictures has done me great good. They have given me comfort. I go away from this place quieted, and much strengthened to do my duty." He had found his own life in Cole's archetypal, artistic depiction, and felt recognized. Someone had seen, and that helped.

Also, according to Hindu teachings, the human life is divided into four stages. In the first, the sole responsibility of youth is to learn; in the second, one becomes a worldly householder active in the affairs of life; in the third, one retires to the forest for a period of meditation; and finally, in the fourth phase, one becomes a wandering mendicant sage.

Six centuries before Thomas Cole, but long after the rise of

Hindu philosophy, an unknown artist painted the crypt ceiling of the Cathedral of Anagni in central Italy, also showing four ages to life's span.[9] Wonderfully geometrical in its composition, a series of concentric circles spreads like ripples in a pond, linking a peripheral cosmos with HOMO, the human being. Shown nude and upright, the human form is ringed around with the words *mikrocosmos, id est minor mundus,* "microcosm, that is to say, a little world." Man is a little world. The rings are segmented neatly into four quadrants, and in each is the painted bust of an age of man: child, youth, prime, old age. With each is also written the corresponding psychological temperament: sanguine, choleric, melancholic, and phlegmatic, respectively. In every nuance of the painting, an ordering of both human life and the natural universe announces itself.

Absent in Anagni is Cole's powerful, dynamic play of light and color; the figures are motionless models from whom it is difficult to imagine much comfort could be drawn for a troubled life. Each frame is complete unto itself. These are static stages of life, not moments in a passage through it. But whether in Anagni, in India, or Utica, New York, the stations of life's journey are objects of meditation, and these are, most notably, inner, developmental stages whose character implicitly shapes the character of our knowing, not an arid outer chronology. I have tried to paint the history of light in similar colors, from the inside as well as the outside.

Human history and biography are inevitably suffused with the operations of the psyche. Only more recently have we grown certain of a similar entanglement of outer nature and the human mind. The transformations of cultures over time have had profound effects on the insights humanity has had into nature. We have seen the character of successive ages reflected in the images they have made of light. These form a sequence, not of disjointed fragments, but a whole that unfolds in time, a series of awakenings that bespeaks an inner evolutionary development. Heraclitus was right: "It is in changing that things find repose."

What seems eternal, must be seen anew: "The sun is new each day." Scientific and prescientific knowledge of light have reflected the continual metamorphosis of our inward instruments of knowing. The very existence of that transformation suggests the possibility of further evolution, individually and culturally, and the possibility of relinking the moral and sensual, the physical and spiritual, in a fresh, unitary imagination.

Past change occurred with little self-consciousness. Mistakes could be left behind. The time of unconscious change is over, as environmental and nuclear hazards daily bring home to us. We now inhabit the entire planet, and have learned the potency of our accomplishments. Future evolution must be shaped self-consciously. It is a new and dangerous technology.

What should be the nature of future knowledge; how will we see light tomorrow? From all that has gone before, a fruitful avenue seems clear. First, it will require the modest recognition that we all are only partly sighted creatures, and so know only a part of nature. As Novalis—who was both a poet and a mining engineer—wrote: "But it is vain to attempt to teach and preach Nature. One born blind does not learn to see though we tell him forever about colors, lights and distant forms. Just so no one will understand Nature who has not the necessary organ, the inward instrument, the specific creating instrument. . . . "[10]

If we lack the "necessary organ, the inward instrument" as Novalis called it, we will need to cultivate it.

Clearsight

Finally, I must tell you that as a painter I am becoming more clearsighted before nature.

Paul Cézanne[11]

Cézanne often painted the same scene over and over again.

Standing on the bank of a river, he told his son that the motifs he saw so multiplied themselves "that I think I could occupy myself for months without changing place." What did Cézanne hope to accomplish, or to see, by tirelessly reworking the same vista? As if by way of answering, Cézanne wrote to Emile Bernard: "Get to the heart of what is before you. . . . In order to make progress, there is only nature, and the eye is trained through contact with her. It becomes concentric through looking and working."[12]

The eye becomes concentric, aligned with nature, through the artist's ceaseless action of looking and working, of struggling to see clearly a single gesture of nature's infinitely varied repertoire, and then to paint it. In guiding the hand across a canvas, one fashions and refines fresh senses, new capacities of mind suited to seeing that which until then had eluded the eye. In the end, one "gets to the heart of what is before us." Like the alchemist, whose outer actions were but an image of an inner transformation, the artist, in creating outwardly, simultaneously accomplishes an equally precious inner work—clearsight.

—

SITTING QUIETLY BENEATH the shade of a tree, a Buddhist monk meditates on one of the forty traditional subjects for meditation that form the basis of Buddhist practice. At knee level before him is the earth *kasina,* a disk of earth alike in all outward respects to a carefully crafted mud pie, and the object of his unswerving contemplation. As he meditates on the earth *kasina* it gradually disappears. In its place hovers an "afterimage" that initially may last only a few minutes, but that in the end becomes as firm in the mind as the mud pie is in the shade of the tree. Following the earth *kasina* further leads the meditant through a series of disappearances and new appearances that, for the Buddhist adept, demarcate transitions to realms beyond everyday consciousness. Over the course of months and years, if necessary, the monk will follow the images of earth from one realm to another until he, like Cézanne, has become sufficiently clearsighted to get to the heart or essence of earth. Then the sign of earth becomes, as Buddhaghosa described it, "like the disc of a mirror, a well-burnished conch-vessel, the round moon issuing from the clouds, white cranes against a rain cloud, and makes its appearance as though bursting the grasped sign, than which it is a hundred times, a thousand times more purified."[13]

Among the nine other *kasinas* or devices (which include water, air, heat, blue, yellow, red, white, and space) is also the *kasina* of light. Of it, Buddhaghosa writes: "he who grasps the Light-device grasps the sign in light entering through a wall-crevice, key-hole, or window-space."[14] That is, every manifestation of light is potentially the occasion for the true grasping of light, be it the dappled disks of light beneath the shade of a tree, or the moonbeam that furtively makes its way through a chink in the wall. Each instance offers an occasion for enlightenment, for seeing light.

William Blake boldly put it this way: "If the doors of perception were cleansed, everything would appear to man as it is, infinite."

—

THE PHILOSOPHER SCHOPENHAUER once recorded a remarkable conversation between Goethe and himself concerning light. Schopenhauer sensibly suggested that light is a purely subjective, psychological phenomenon, and that without sight, light could not be said to exist. Goethe responded vehemently, as Schopenhauer describes it: " 'What,' he [Goethe] once said to me, staring at me with his Jupiter-like eyes. 'Light should only exist in as much as it is seen? No! *You* would not exist if the light did not see *you*!' "[15]

Goethe as botanist well knew the life-giving powers of light.

In addition, he felt that light brought forth not only life but also could, through its ceaseless action, create the very organ suited to perceive it. Evolution has occurred in the context of light, and over time the body responded with the organ of sight. Goethe phrased it this way: "The eye owes its existence to the light. Out of indifferent animal organs the light produces an organ to correspond to itself; and so the eye is formed by the light for the light so that the inner light may meet the outer."[16] Light, ever active, created the eye. It sculpted an organ suited to itself, like the streaming water shaping the stones over and through which it flows. Had light not "seen" man, we should never have seen the light.

What Goethe pointed to at the concrete level of human physiology is likewise true for more subtle organs of soul. Cézanne's looking and working, the Buddhist in his meditations, Goethe through his ceaseless artistic and scientific undertakings, each of these aimed at the birthing of the "necessary organ, the inward instrument" needed to see more deeply into nature. And the opportunity for this practice is ever-present, for, as Goethe says, "Every object, well contemplated, opens up a new organ within us."[17]

The artist and the monk both know that through a disciplined practice they can internalize nature so that they can realize new capacities of mind. Personal growth is not only a matter of memorizing sacred texts (which the monk has certainly done), or of academic artistic analysis (which Cézanne has also done), but it requires praxis, daily labor, to fashion fresh, hard-won, soul faculties. Every action of the hand and the eye sculpts the soul. Piaget called the process accommodation, the development of new cognitive structures. Goethe, Novalis, Emerson, Steiner . . . spoke of new senses that would reveal fresh aspects of nature's infinite being.

The artist and monk are distinct from us not because of what takes place in them, but because they apply themselves self-consciously to transformation. They educate themselves for an end they have chosen. By contrast, most of us are educated by others and for ends we have not chosen. Traditionally, artists, philosophers, and religious figures have formed that small, self-conscious segment of society that engages in the important if often painful business of introspection and prophetic critique. They sense the dangers of an unthinking, habitual mode of seeing and know the need for tireless renewal.

—

IF EVERY CULTURE and period have thought so variously about light, and if quantum theory has stripped light of its naively imagined attributes, then what of certainty is left to the nature of light at the end of this evolutionary drama? Everything. The true artist, monk, and scientist are not searching to grasp knowledge as object, but rather as event. The moment of critical importance is the moment pointed to by Goethe, the moment of aperçu, or insight. For millennia one can see the sun rise and never notice the rotation of the earth. One can throw a thousand rocks and never see their parabolic flight. We can wake each morning for sixty years to the glow of the dawn and never see light. Why? Because one rushes past the immediate offerings of the senses to what we suppose to be the hidden, enduring, primary objects of reality. The habits of our culture, the dogmas of our education constrain our sight. Atoms become immortal gods, photons their stern messengers.

Knowledge is not an object to be traded like chattel, but an epiphanous moment to be cherished. Too often we bypass epiphanies for a currency more ready to hand, for an abstract notion, an old insight in new dress, an equation whose characters could reveal, but which, in fact, we allow to go unread. These are the stones of knowing instead of the bread, they are idols instead of gods. Epiphanous knowing presupposes organs for insight, inward instruments; and new knowing requires new instruments.

We each possess the rudiments of every organ, but we deny them the nurture they need, we neglect the praxis in whose light they might grow and flourish.

The photon is not an object to be hefted in the hand like a clod of earth. Its very elusiveness draws us to the essentials of knowing as seeing, to knowledge as event. Understood in this way, insights into light's multifaceted nature are not the exclusive property of twentieth-century physics. Our present insights do not exhaust the wealth of light, but complement the genuine insights of the past. The future also becomes clearer. The archetypal experiments of quantum optics and the meditations of the contemporary poet offer us equally the opportunity for fresh aperçu, but only if we have the patience to educate ourselves in light, to make ourselves concentric with its nature, as Cézanne and Buddhaghosa recommend.

—

OVER MILLENNIA, CULTURES have embraced and discarded countless images of light. Within a single lifetime, likewise, we have lived within and shed successive understandings of light. Through research, artistic praxis, and quiet contemplation, light's elusive being constantly re-creates itself in our mind's eye, offering fresh epiphanies to every generation. When seen with a thousand eyes, light will finally rest with us in the haven we have made.

Seeing light is a metaphor for seeing the invisible in the visible, for detecting the fragile imaginal garment that holds our planet and all existence together. Once we have learned to see light, surely everything else will follow.

Notes

Chapter 1 / Entwined Lights:
The Lights of Nature and of Mind

1. Lao-tzu, *Tao Te Ching*, quoted in Hastings, *Encyclopedia of Religion* (New York: Charles Scribner's Sons, 1908–26), vol. 8, p. 51.
2. Quoted in M. von Senden, *Space and Sight: The Perception of Space and Shape in the Congenitally Blind before and after Operation*, trans. Peter Heath (Glencoe, IL: The Free Press, 1960), p. 40.
3. R. L. Gregory and J. G. Wallace, "Recovery from Early Blindness: A Case Study," in *Perception*, ed. Paul Tibbetts (New York: Quadrangle/New York Times Book Co., 1969).
4. M. von Senden, *Space and Sight*, p. 20.
5. David H. Hubel, *Eye, Brain and Vision* (New York: Scientific American Library, distributed by W. H. Freeman, 1988), chap. 9.
6. Quoted in M. von Senden, *Space and Sight*, p. 160.
7. Marjorie Hope Nicolson, *Newton Demands the Muse* (Princeton, NJ: Princeton University Press, 1966), p. 75.

Chapter 2 / The Gift of Light

1. Plato, "Protagoras," sections 320d–321e. *The Collected Dialogues of Plato*, ed. Edith Hamilton and Huntington Cairns (Princeton, NJ: Princeton University Press, 1969).
2. Hesiod, *Works and Days*, trans. Apostolos N. Athanassakis (Baltimore: Johns Hopkins University Press, 1983), lines 50–100.
3. Plato, "Symposium," 219a.
4. Homer, *The Odyssey*, trans. Robert Fitzgerald (Garden City, NY: Doubleday, 1961), bk. 3, lines 1–4.
5. Eleanor Irwin, *Color Terms in Greek Poetry* (Toronto: Hakkert, 1974).
6. Homer, *The Iliad*, trans. Robert Fitzgerald (Garden City, NY: Doubleday, 1974), bk. 24, lines 401–03.
7. Homer, *Iliad*, bk. 24, lines 93–94.
8. Oliver Sacks and Robert Wasserman, "The Case of the Colorblind Painter," *The New York Review of Books*, vol. 34 (November 19, 1987), pp. 25–34.
9. Benjamin Whorf, *Language, Thought and Reality* (Cambridge: MIT Press, 1964).
10. Diogenes Laertius, *Lives of the Philosophers*, ed. and trans. A. Robert Caponigri (Chicago: Henry Regnery Co., 1969), chap. 8.
11. See E. R. Dodds, *The Greeks and the Irrational* (Berkeley, CA: University of California Press, 1951), pp. 145–46.
12. Empedocles, *Katharmoi (Purifications)* in Kathleen Freeman, *Ancilla to the Pre-Socratic Philosophers* (Cambridge: Harvard University Press, 1983), p. 65, frag. 115.
13. Empedocles, *On Nature*, in Freeman, *Ancilla*, p. 61, frags. 85–87.
14. Freeman, *Ancilla*, pp. 60–61, frag. 84.
15. The Revised Standard Version of the Bible translates Matthew 6:22–23 as "The eye is the lamp of the body. So if

your eye is sound, your whole body will be full of light; but if your eye is not sound, your whole body will be full of darkness. If then the light in you is darkness, how great is the darkness."

16. Freeman, *Ancilla,* p. 58, frag. 48.

17. Plato, *Timaeus,* 45b; David C. Lindberg, *Theories of Vision from al-Kindi to Kepler* (Chicago: University of Chicago Press, 1976), chap. 1.

18. This is a paraphrase of a Neo-platonic text. See J. W. von Goethe, *Theory of Color* in *Scientific Studies,* ed. and trans. Douglas Miller, vol. 12 of *Goethe: Collected Works in English* (New York: Suhrkamp, 1988), p. 164.

19. Paul Friedlaender, *Plato, an Introduction,* trans. Hans Meyerhoff (Princeton, NJ: Bollingen, 1973), p. 13.

20. De Lacy Evans O'Leary, *How Greek Science Passed to the Arabs* (London: Routledge & Kegan Paul, 1964).

21. On Ibn al-Haytham or Alhazen see the article by A. I. Sabra in *The Dictionary of Scientific Biography,* ed. Charles Gillispie (New York: Charles Scribner's Sons, 1972), vol. 6, pp. 189–210; and Lindberg, *Theories of Vision,* pp. 60–86.

22. Quoted in *Dictionary of Scientific Biography,* vol. 6, p. 190.

23. Lindberg, *Theories of Vision,* p. 91.

24. A. C. Crombie, "Early Concepts of the Senses and the Mind," *Scientific American,* vol. 210, no. 5 (May 1964), pp. 108–16; see also Lindberg's objections to Crombie in *Theories of Vision,* p. 207.

25. Lindberg, *Theories of Vision,* p. 66.

26. The *camera obscura* was probably known in some form in antiquity. References in Euclid and the pseudo-Aristotle *Problemata* indicate some familiarity with it, but Alhazen appears to be the first to describe it fully. See John Hammond, *The Camera Obscura: A Chronicle* (Bristol, CT: Hilger, 1987), pp. 1–7.

27. Johannes Kepler, quoted in Lindberg, *Theories of Vision,* p. 203. Kepler combined his work with Felix Plater's recognition of the retina (instead of the lens) as the sensitive tissue of the eye.
28. Lindberg, *Theories of Vision,* p. 203.
29. Lindberg, *Theories of Vision,* p. 201.
30. Hubel, *Eye, Brain, and Vision,* p. 222.
31. Owen Barfield, *Saving the Appearances* (New York: Harcourt, Brace & World, 1965).

Chapter 3 / Light Divided: Divine Light and Optical Science

1. C. S. Lewis, *The Discarded Image* (Cambridge: Cambridge University Press, 1964), pp. 222–23.
2. Ahura Mazda in Persian mythology has the sun for his eye. See M. N. Dhalla, *History of Zoroastrianism* (New York: Oxford University Press, 1938), p. 213.
3. W. Max Mueller, *Egyptian Mythology* in *The Mythology of All Races,* ed. Louis Herbert Gray (Boston: Marshall Jones Co., 1943), vol. 12, p. 70.
4. Veronica Ions, *Egyptian Mythology* (New York: Paul Hamlyn, 1975), p. 41.
5. Mary Boyce, ed. and trans., *Textual Sources for the Study of Zoroastrianism* (Totowa, NJ: Barnes and Noble Books, 1984), p. 75. Some authors, ancient and modern, put Zoroaster back as far as 6000 B.C.
6. Mary Boyce, *Zoroastrians: Their Religious Beliefs and Practices* (London: Routledge & Kegan Paul, 1979); and Boyce, *Textual Sources for the Study of Zoroastrianism.*
7. RSV, Genesis 1:14–19.
8. RSV, Genesis 3:16–19.
9. Jeffery Burton Russell, *The Devil: Perceptions of Evil from*

Antiquity to Primitive Christianity (Ithaca, NY: Cornell University Press, 1978), pp. 207–09.

10. RSV, 1 John 1:5.

11. RSV, John 1:67–69.

12. Jes P. Asmussen, *Manichaean Literature* (Delmar, NY: Scholars' Facsimiles & Reprints, 1975), p. 88.

13. Samuel N. C. Lieu, *Manichaeism in the Later Roman Empire and Medieval China* (Manchester, U.K.: Manchester University Press, 1985), pp. 210–13.

14. *The Cologne Mani Codex, "Concerning the Origin of his Body,"* ed. and trans. Ron Cameron and Arthur J. Dewey (Missoula, MT: Scholars Press, 1979), p. 19.

15. G. Van Groningen, *First Century Gnosticism: Its Origins and Motifs* (Leiden, Neth.: E. J. Brill, 1967), pp. 23ff.

16. "Two Cathars Tales," *Journal for Anthroposophy*, trans. Christopher Bamford, no. 36 (Autumn 1982).

17. Zoe Oldenbourg, *Massacre at Montsegur*, trans. by Peter Green (London: Wiedenfeld & Nicolson, 1961), p. 50.

18. Steven Runciman, *The Medieval Manichee* (Cambridge: Cambridge University Press, 1982).

19. Denis de Rougemont, *Love in the Western World* (Princeton, NJ: Princeton University Press, 1983), p. 82.

20. James McEvoy, *The Philosophy of Robert Grosseteste* (Oxford: Clarendon Press, 1982).

21. Alexandre Koyré, *Diogenes* 4 (1957), pp. 421–48; and McEvoy, *The Philosophy of Robert Grosseteste*, p. 210.

22. For the importance of light and measure in Gothic architecture see Otto von Simson, *The Gothic Cathedral* (Princeton, NJ: Princeton University Press, 1988), chapter 2.

23. The Bible, Wisdom 11:21; this was a favorite text of Grosseteste's. On Grosseteste's view of God as mathematician, see McEvoy, pp. 167–80.

24. Louis Kahn, interviewed in *Time*, January 15, 1973.

25. Robert Grosseteste, quoted in A. C. Crombie, *Robert*

Grosseteste and the Origins of Experimental Science, 100–1700 (Oxford: Clarendon Press, 1953), p. 143.

26. Crombie, *Experimental Science*, pp. 10–11.
27. See Lindberg, *Theories of Vision*, chap. 2.
28. McEvoy, *Philosophy of Robert Grosseteste*, p. 372.

Chapter 4 / The Anatomy of Light

1. Samuel Edgerton, Jr., *The Renaissance Rediscovery of Linear Perspective* (New York: Harper & Row, 1976), especially chaps. 1 and 10. Also see John White, *The Birth and Rebirth of Pictorial Space* (London: Faber & Faber, 1967).
2. Martin Kemp, *The Science of Art* (New Haven: Yale University Press, 1990), p. 9.
3. Antonio di Tuccio Manetti, *The Life of Brunelleschi*, trans. Catherine Enggass (University Park, PA: Pennsylvania State University Press, 1970), pp. 42–46.
4. Ernst Cassirer, *Language and Myth*, trans. Suzanne K. Langer (New York: Harper & Brothers, 1946), p. 8.
5. Jan Deregowski, "Pictorial Perception and Culture," reprinted in *Image, Object and Illusion* (San Francisco: W. H. Freeman & Company, 1974), chap. 8.
6. Suzi Gablik, *Progress in Art* (New York: Rizzoli, 1976), chaps. 6–7.
7. Edgerton, *The Renaissance Rediscovery of Linear Perspective*, chap. 10, pp. 157–58, summarizing the article by Erwin Panofsky, "Die Perspecktive als 'symbolische Form,' " *Vortraege der Bibliotek Warburg 1924–25* (Leipzig, Ger.: 1927), pp. 258–331.
8. Quoted by Richard Krautheimer and Trude Krautheimer-Hess, *Lorenzo Ghiberti* (Princeton, NJ: Princeton University Press, 1956), p. 14.

9. Leonardo da Vinci, *The Notebooks of Leonardo da Vinci*, ed. and trans. Edward MacCurdy (New York: George Braziller, 1955), p. 93.

10. Otto von Simson, *The Gothic Cathedral* (Princeton, NJ: Princeton University Press, 1988), chap. 2.

11. Ernst Cassirer, *The Individual and the Cosmos in Renaissance Philosophy*, trans. by Mario Domandi (Philadelphia: University of Pennsylvania Press, 1963), chaps. 1 and 2.

12. Leonardo, *Notebooks*, p. 57.

13. Leonardo, *Notebooks*, p. 612.

14. See the Masonic Temple Legend for interesting background in Charles William Heckethorn, *Secret Societies* (New Hyde Park, NY: University Books, 1965), vol. 1, bk. 8, chap. 1.

15. George Ovitt, Jr., *The Restoration of Perfection* (New Brunswick, NJ: Rutgers University Press, 1987); and Lynn White, Jr., *Medieval Technology and Social Change* (New York: Oxford Galaxy Book, 1966).

16. Galileo Galilei, *Dialogue*, trans. Stillman Drake (Berkeley, CA: University of California Press, 1953).

17. Stillman Drake, *Galileo* (New York: Hill & Wang, 1980), p. 24.

18. Galileo Galilei, "Letter to the Grand Duchess Christina," *Discoveries and Opinions of Galileo*, trans. Stillman Drake (Garden City, NY: Doubleday & Co., 1957), p. 182.

19. Homer, *Odyssey*, bk. 6, line 243.

20. Homer, *Odyssey*, bk. 8. Odysseus, modern soul that he was, warned that looks can be deceiving: "In looks a man may be a shade, a specter, and yet be master of speech so crowned with beauty that people gaze at him with pleasure."

21. Johannes Kepler, *Epitome of Copernican Astronomy*, bk. IV, I, 3.

22. Plato, *Laws*, bk. VII; see also Francis Cornford, *Before and*

After Socrates (Cambridge: Cambridge University Press, 1972).

23. Galileo Galilei, *The Assayer* in *Discoveries and Opinions*, p. 274.

24. The technology is slightly more complicated, relying on a change in the polarization of the light as it passes through the liquid crystal.

25. We would call it barium sulfide.

26. Galileo Galilei, *The Assayer* in *Discoveries and Opinions*, p. 278.

27. Richard S. Westfall, *Never at Rest: A Biography of Isaac Newton* (Cambridge: Cambridge University Press, 1980), pp. 232ff.

28. Westfall, *Never at Rest*, p. 141.

29. Westfall, *Never at Rest*, p. 154.

30. Westfall, *Never at Rest*, p. 426.

31. Rudolf Steiner, *Philosophie und Anthroposophie*, "Mathematik und Okultismus" (Dornach, Switz.: Verlag der Rudolf Steiner-Nachlassverwaltung, 1965).

32. Alan E. Shapiro, "Newton's Definition of a Light Ray," *Isis*, vol. 66 (1975), pp. 194–210.

33. Isaac Newton, *Opticks*, 4th edition (New York: Dover, 1952), p. 1.

34. A. I. Sabra, *Theories of Light from Descartes to Newton* (New York: Cambridge University Press, 1981), chap. 11, for example, argues a dogmatic atomism of light.

35. Shapiro, "Newton's Definition of a Light Ray."

36. G. N. Cantor, *Optics After Newton* (Dover, NH: Manchester University Press, 1983), chap. 2.

37. Newton to Robert Hooke, letter of February 5, 1676, quoted in Westfall, *Never at Rest*, p. 274.

38. Marjorie Hope Nicolson, *Newton Demands the Muse* (Princeton, NJ: Princeton University Press, 1946).

39. John Hughes, *The Ecstasy* (1735), quoted in Nicolson, *Newton Demands the Muse*, p. 11.

40. See, for example, "To the Memory of Sir Isaac Newton," in *The Complete Poetical Works of James Thomson*, ed. J. L. Robertson (London: 1908); or Nicolson, *Newton Demands the Muse*, p. 12.
41. From *Journal de Trevoux* (Paris: Chez E. Ganeau, 1701).
42. Quoted in Stephen Mason, *A History of the Sciences* (New York: Collier Books, 1962), p. 169.
43. Jacques Maritain, *The Dream of Descartes* (New York: Philosophical Library, 1944). See chap. 1 for an engaging account of the dream.
44. Francis A. Yates, *The Rosicrucian Enlightenment* (Boulder, CO: Shambhala, 1978).
45. Quoted in Mason, *History of the Sciences*, p. 169.
46. A. I. Sabra, *Theories of Light*, p. 48.
47. Cantor, *Optics After Newton*, chap. 3.
48. Cantor, *Optics After Newton*, p. 33.
49. Sabra, quoting Descartes; *Theories of Light*, p. 60.
50. Bernard de Fontenelle, *Conversations on the Plurality of Worlds*, trans. H. A. Hargreaves (Berkeley, CA: University of California Press, 1990), "The First Evening."

Chapter 5 / The Singing Flame:
Light as Ethereal Wave

1. Edward Grant, ed., *A Source Book in Medieval Science* (Cambridge: Harvard University Press, 1974). See Bartholomew the Englishman "Concerning the Properties of Things," trans. Bruce Eastwood, p. 383; also Vasco Ronchi, *The Nature of Light*, trans. V. Varocas (Cambridge: Harvard University Press, 1970), p. 62; and Lindberg, *Theories of Vision*, p. 134.
2. Leonardo da Vinci is here quoting John Pecham.
3. René Descartes, in Sambursky's *An Anthology of Physical Thought* (New York: Pica Press, 1975), p. 244.

4. Leonhard Euler, *Letters of Euler on Natural Philosophy Addressed to a German Princess*, ed. David Brewster (New York: J. & J. Harper, 1833), p. 77.

5. Euler, *Letters to a German Princess*, p. 85. The notion of light as vibration existed before Euler, most especially in the thought of Christian Huygens, but it did not successfully challenge Newton's corpuscular view.

6. Adolf Katzenellenbogen, *The Sculptural Programs of Chartres Cathedral* (Baltimore: Johns Hopkins, The University Press, 1959), pp. 15–22.

7. John Hawkins, *General History of the Science and Practice of Music* (New York: Dover, 1963), vol. II, chap. 133. Joscelyn Godwin, *Harmonies of Heaven and Earth* (Rochester, VT: Inner Traditions, 1987).

8. Francis Bacon, quoted in Dayton C. Miller, *Anecdotal History of the Science of Sound* (New York: Macmillan, 1935), p. vi.

9. Aristotle, *Sound and Hearing*, quoted in Miller, *Anecdotal History of Sound*, p. 3.

10. John Leconte, "On the Influence of Musical Sounds on the Flame of a Jet of Coal-Gas," *Philosophical Magazine*, 4th series, vol. 15 (1858), p. 235.

11. Maeterlinck quoted in Arthur Symons, *The Symbolist Movement in Literature* (New York: E. P. Dutton & Co., 1919), p. 92.

12. Adelard of Bath, "Natural Questions," in *A Source Book in Medieval Science*, ed. Edward Grant.

13. Alexander Wood, *Thomas Young, Natural Philosopher (1773–1829)* (Cambridge: Cambridge University Press, 1954); *Dictionary of Scientific Biography*, vol. 14, pp. 562–72.

14. Cantor, *Optics After Newton*, p. 146.

15. Otto Neugebauer, *The Exact Sciences in Antiquity* (Providence, RI: Brown University Press, 1957).

16. R. J. Gillings, "The So-called Euler-Diderot Incident," *American Mathematical Monthly*, vol. 61 (1954), pp. 77–80.

17. François Arago, *Biographies of Distinguished Scientific Men*, trans. Smyth, Powell, and Grant (London: Longman, 1857), pp. 399–471.

18. Robert Greenler, *Rainbows, Halos, and Glories* (New York: Cambridge University Press, 1980), chap. 6.

19. Sound waves are "longitudinal waves," which means that the vibration is in the same direction as the direction of propagation and not transverse to it.

20. Sir George Stokes, "On the Constitution of the Ether," May 1848, in *Nineteenth-Century Aether Theories*, ed. Kenneth F. Schaffner (New York: Pergamon Press, 1972).

21. G. N. Cantor, "The Theological Significance of Ethers," *Conceptions of Ether*, eds. G. N. Cantor and M.J.S. Hodge (Cambridge: Cambridge University Press, 1981), p. 149.

22. Spiritualism sought for physically demonstrable effects of the spirit, and so materialized psychical research and experimentation. It clearly remained true to the materialistic imagination of the nineteenth century, even when treating the immaterial.

Chapter 6 / Radiant Fields:
Seeing by the Light of Electricity

1. Joseph Agassi, *Faraday as a Natural Philosopher* (Chicago: University of Chicago Press, 1971); and L. Pearce Williams, *Michael Faraday* (New York: Basic Books, 1965). I have relied heavily on L. Pearce Williams's biography of Faraday throughout this chapter.

2. John Tyndall, *Faraday as a Discoverer* (New York: D. Appleton and Co., 1873).

3. Quoted by G. N. Cantor, "Reading the Book of Nature: The

Relation Between Faraday's Religion and His Science," in *Faraday Rediscovered* (New York: Stockton Press, 1985), eds. David Gooding and Frank James, p. 71.

4. Tyndall (1894), quoted by Cantor, "Reading the Book of Nature," p. 74.

5. Cantor, "Reading the Book of Nature," p. 74.

6. Michael Faraday, *Experimental Researches in Electricity* (London: Richard Taylor & William Francis, 1855), vol. 2, pp. 284ff.

7. See for example the account of Frank A.J.L. James, " 'The Optical Mode of Investigation': Light and Matter in Faraday's Natural Philosophy," in *Faraday Rediscovered*, p. 149.

8. Faraday, *Experimental Researches*, vol. 3, pp. 447ff.

9. See for example Faraday's June 1852 paper, "On the Physical Lines of Magnetic Force," in *Experimental Researches*, vol. 3, pp. 438ff.

10. Faraday, *Experimental Researches*, vol. 2, p. 451.

11. Some scholars argue that Faraday advanced in turn two types of field theories. One operated through empty space by the very properties of the field or space itself (his 1846 view), and later, perhaps under the influence of W. Thompson (Kelvin), Faraday changed his view completely to that of a continuous, nonparticulate ether. (See Barbara Giusti Doran, "Origins and Consolidation of Field Theory in Nineteenth-Century Britain," in *Historical Studies in the Physical Sciences*, [Princeton, NJ: Princeton University Press, 1975], vol. 6, pp. 133–260.) I disagree. From reading the sources, I suspect that the second view was never really adopted by Faraday. When he does mention the ether after 1846, he usually surrounds the reference with so many qualifiers that he effectively keeps it at arm's length, while accommodating the enthusiasms of colleagues such as Kelvin.

12. Faraday, *Experimental Researches*, vol. 2, p. 452.
13. Boethius, *The Consolation of Philosophy*, trans. V. E. Watts (Baltimore: Penguin, 1969).
14. Alan of Lille, *Plaint of Nature*, trans. James J. Sheridan (Toronto: Pontifical Institute, 1980).
15. See the "Night" scene in Johann Wolfgang von Goethe, *Faust*, ed. and trans. Stuart Atkins (Boston: Suhrkamp/ Insel, 1984).
16. See "A Vision" in the poems of James Clerk Maxwell printed at the end of *The Life of James Clerk Maxwell*, Lewis Campbell and William Garnet (London: Macmillan and Co., 1882).
17. Sir Thomas Browne, *Religio Medici and Other Works*, ed. L. C. Martin (Oxford: Clarendon Press, 1964).
18. Quoted in Ivan Tolstoy, *James Clerk Maxwell* (Edinburgh: Canongate, 1981), p. 59.
19. Friedrich Hölderlin, *Brot und Wein* in *Poems and Fragments*, trans. Michael Hamburger (Cambridge: Cambridge University Press, 1980).
20. Albert Einstein, "Maxwell's Influence on the Development of the Conception of Physical Reality," in Sir J. J. Thomson, *James Clerk Maxwell* (New York: Macmillan, 1931), pp. 66–67.
21. Richard Feynman, *Lectures on Physics* (Reading, MA: Addison-Wesley, 1965), vol. 2.
22. Jed Z. Buchwald, *From Maxwell to Microphysics: Aspects of Electromagnetic Theory in the Last Quarter of the Nineteenth Century* (Chicago: University of Chicago Press, 1985).
23. On William Thompson, Lord Kelvin, see Harold Issadore Sharlin, *Lord Kelvin: The Dynamic Victorian* (University Park, PA: Pennsylvania State University Press, 1979), and David B. Wilson, *Kelvin and Stokes* (Bristol, CT: Adam Hilger, 1987).

24. Willard Gibbs, quoted in *Dictionary of Scientific Biography*, vol. 13, p. 386.
25. Sharlin, *Lord Kelvin*, p. 237.
26. Quoted in Sharlin, *Lord Kelvin*, p. 226.
27. Tyndall, *Faraday as a Discoverer*, p. xvii.
28. John Stuart Mill, *Autobiography*, 1924, p. 129. Quoted in "The 'Spectre' of Science," by C. J. Wright, *Journal of the Warburg and Courtauld Institutes* (1980), vol. 43, p. 187.
29. Henry David Thoreau, *The Journals of Henry D. Thoreau*, ed. Francis H. Allen and Bradford Torrey (Boston: Houghton Mifflin Co., 1906), vol. III, pp. 155–56.
30. Quoted in Keiji Nishitani, *Religion and Nothingness*, trans. Jan Van Bragt (Berkeley, CA: University of California Press, 1982), p. 167.
31. RSV, Genesis 9:13.

Chapter 7 / Door of the Rainbow

1. Black Elk, *Black Elk Speaks* (Lincoln, NE: University of Nebraska Press, 1988), p. 25.
2. Homer, *Iliad*, bk. 2.
3. Flood myths are extraordinarily widespread, ranging from Plato's story of the lost continent of Atlantis to South American Indian legends. For further discussion see *The Flood Myth*, ed. Alan Dundes (Berkeley, CA: University of California Press, 1988).
4. Plato, *Theatetus*, 155d.
5. Xenophanes, frag. 28D.
6. December 10, 1834, off Chonos Archipelago. Charles Darwin, *Journal of Researches during the Voyage of H.M.S. Beagle* (New York: Hafner Publishing Co., 1952), p. 269.
7. Aristotle, *Meteorologica*, trans. H. D. P. Lee (Cambridge: Harvard University Press, 1952), bk. II, 371b, line 19.
8. Carl B. Boyer, *The Rainbow: From Myth to Mathematics*

(Princeton, NJ: Princeton University Press, 1989); Robert Greenler, *Rainbows, Halos, and Glories.*

9. Boyer, *The Rainbow*, pp. 89ff.; a translation of *De Iride* is contained in Bruce S. Eastwood, "Robert Grosseteste on Refraction Phenomena," *American Journal of Physics*, vol. 38, pp. 196ff (1970).

10. Grosseteste, in Eastwood, *American Journal of Physics*, p. 196.

11. Eastwood, *American Journal of Physics*, p. 198.

12. Theodoric of Freiberg, quoted in Boyer, *The Rainbow*, p. 114.

13. Boyer, *The Rainbow*, p. 116.

14. James Thomson, *The Complete Poetical Works of James Thomson*, ed. J. L. Robertson (New York: 1908), pp. 436–42.

15. See "Newton's Rainbow and the Poet's" in M. H. Abrams, *The Mirror and the Lamp* (Oxford: Oxford University Press, 1953).

16. *The Autobiography and Memoirs of Benjamin Haydon*, edited from his Journal by Tom Taylor, a new edition with an introduction by Aldous Huxley (New York: Harcourt, Brace, 1926), vol. 1, p. 269.

17. Thomas Campbell, "The Rainbow" (1820).

18. Lao-tzu, *Tao Te Ching*, trans. Stephen Mitchell (New York: Harper & Row, 1988), p. 52.

19. Owen Barfield, *Saving the Appearances: A Study in Idolatry* (New York: Harcourt, Brace & World, 1965).

20. J. W. von Goethe, *Zur Farbenlehre*, in *Goethes Werke, Hamburger Ausgabe*, ed. Erich Trunz (Munich, Ger.: C. H. Beck, 1982), vol. 13, my translation. See also J. W. von Goethe, "Introduction to *Theory of Color*" in *Scientific Studies*, p. 164.

21. Michael Polanyi, *Knowing and Being* (Chicago: University of Chicago Press, 1969), p. 148.

22. John Stuart Mill, *Mill on Bentham and Coleridge*, ed. F. R. Leavis (London: Chatto & Windus, 1950), pp. 99–100.

23. J. W. von Goethe, *Faust*, trans. Albert G. Latham (New York: E. P. Dutton, 1908), pt. II, p. 15.

Chapter 8 / Seeing Light—Ensouling Science: Goethe and Steiner

1. Edwin Land, *Proceeding of the National Academy of Sciences*, vol. 45 (1959), pp. 115–29, 636–44; and *Scientific American*, vol. 200 (May 1959), pp. 84–99.

2. Michael Wilson and R. W. Brocklebank, *The Journal of Photographic Science*, vol. 8 (1960), pp. 141–50, and *Contemporary Physics*, vol. 3 (1961), pp. 91–111.

3. J. W. von Goethe, from a letter to the artist Josef Carl Stieler, in *Goethe's Color Theory*, arr. and ed. by Rupprecht Matthaei, trans. by Herb Aach (New York: Van Nostrand Reinhold, 1971), p. 202.

4. This episode is recounted in "Confessions of the Author," included in *Goethe's Color Theory*, p. 199.

5. J. P. Eckermann, *Conversations with Goethe*, trans. Gisela C. O'Brien (New York: Frederick Ungar, 1964), Thursday, February 19, 1829, p. 149.

6. Goethe, *Scientific Studies*, pp. 168ff.

7. Discussed in Rudolf Magnus, *Goethe as Scientist*, trans. Heinz Norden (New York: Schuman, 1949).

8. In recording a conversation with Goethe on February 1, 1827, Eckermann states: "This led our conversation to a great law which pervades all nature, and on which all life and all joy depend. 'This [after-imaging],' said Goethe, 'is the case not only with all our other senses, but also with our higher spiritual nature; and it is because the eye is so eminent a sense, that this law of required change is so striking and so especially clear with respect to colors.'"

J. P. Eckermann, *Conversations of Goethe with Eckermann and Soret*, trans. J. Oxenford (London: Bell, 1883), rev. ed., pp. 216–17.

9. Goethe, *Scientific Studies*, p. 178.

10. The easiest arrangement is to use the light from two flashlights (or candles), one colored, for example, with a red cellophane and the other left uncolored (this light should be dimmer if possible). Set them apart so two clear shadows can be formed. One will appear red and the other green. I sometimes see colored shadows in the evening when the red of the sunset passes through a window and mingles with the white light of a room light. Two distinctly colored shadows appear.

11. Michael Wilson, manuscript report on "Color in Therapy" presented to the Color Group of Great Britain, Imperial College, February 3, 1971.

12. Goethe, quoted in Denis L. Sepper, *Goethe Contra Newton* (Cambridge: Cambridge University Press, 1988), p. 28.

13. J. W. von Goethe, "Konfession des Verfassers," *Goethes Werke, Hamburger Ausgabe*, vol. 14, pp. 251–69; and Sepper, *Goethe Contra Newton*, chap. 2.

14. Goethe, "Konfession des Verfassers," p. 254.

15. G. W. F. Hegel, *Philosophy of Nature*, ed. and trans. M. J. Petry (London: Allen, 1970), vol. 2, sec. 276, p. 17.

16. Goethe, *Scientific Studies*, p. 158.

17. Goethe, *Scientific Studies*, p. 158.

18. Goethe, *Scientific Studies*, p. 307. This excerpt only retranslated by Frederick Amrine.

19. Quoted by Benjamin DeMott, in *Teaching What We Do* (Amherst, MA: Amherst College Press, 1991)

20. J. W. von Goethe, "Significant Help Given by an Ingenious Turn of Phrase" in *Scientific Studies*, pp. 39ff.

21. Goethe, *Scientific Studies*, p. 164.

22. Goethe, *Scientific Studies*, p. 39.

23. J. W. von Goethe, "Empirical Observation and Science," a letter to Schiller, January 15, 1798 in *Scientific Studies*, pp. 24–25.

24. Goethe, *Scientific Studies*, p. 158. This famous line has been translated innumerable ways.

25. Goethe, *Scientific Studies*, p. 307.

26. Goethe, *Scientific Studies*, p. 21.

27. Goethe, in J. P. Eckermann, *Conversations with Goethe*, February 18, 1829, p. 147.

28. Gaston Bachelard, *The Flame of a Candle*, trans. Joni Caldwell (Dallas, TX: Dallas Institute of Humanities and Culture, 1988), p. 20.

29. *Lexikon der Goethe Zitate*, ed. Richard Dobel (Zurich: Artemis Verlag, 1968), p. 524, no. 18.

30. Friedrich von Müller, in *Lexikon der Goethe Zitate*, p. 534, no. 25.

31. Strakosch quoted in Sixten Ringbom, *The Sounding Cosmos* (Abo, Fin.: Abo Akademi, 1970), p. 67.

32. Rudolf Steiner, *Philosophy of Spiritual Activity*, trans. William Lindeman (Hudson, NY: Anthroposophic Press, 1986).

33. Rudolf Steiner, *The Course of My Life*, trans. Olin Wannamaker (Hudson, NY: Anthroposophic Press, 1951), p. 69.

34. Rudolf Steiner, *Riddle of Man*, trans. William Lindeman (Spring Valley, NY: Mercury Press, 1990), p. 130.

35. Rudolf Steiner, *Truth-Wrought Words*, trans. Arvia MacKaye Ege (Spring Valley, NY: Anthroposophic Press, 1979), p. 184.

36. Rudolf Steiner, *The Gospel of John*, trans. Maud Monges (Hudson, NY: Anthroposophic Press, 1984), lecture of May 20, 1908.

37. Rudolf Steiner, "The Connection of the Natural with the Moral-Physical. Living in Light and Weight," in *Colour* (London: Rudolf Steiner Press, 1935), pp. 87ff.

38. Rudolf Steiner, *Truth-Wrought Words,* p. 185.
39. Ralph Waldo Emerson, "Nature," chap. 4, *The Selected Writings of Ralph Waldo Emerson,* ed. Brooks Atkinson (New York: Modern Library, 1968), pp. 15, 18.
40. Rudolf Steiner, *Fruits of Anthroposophy* (London: Rudolf Steiner Press, 1986), pp. 48–49.

Chapter 9 / Quantum Theory by Candlelight

1. Michael Faraday, *The Chemical History of a Candle* (New York: Viking Press, 1960).
2. Gaston Bachelard, *The Flame of a Candle,* p. 20.
3. Paul Claudel, *The Eye Listens,* trans. Elsie Pell (New York: Philosophical Library, 1950), p. 154.
4. Novalis, *The Novices of Sais,* trans. R. Manheim (New York: C. Valentine, 1949), p. 95.
5. Thomas Kuhn, *Black-Body Theory and the Quantum Discontinuity 1894–1912* (New York: Oxford University Press, 1978). An excellent reference for much of the material concerning black-body radiation, Max Planck, and the discovery of the quantum of action.
6. J. L. Heilbron, *The Dilemmas of an Upright Man* (Berkeley, CA: University of California Press, 1986).
7. A. Einstein, "On a Heuristic Point of View Concerning the Production and Transformation of Light," *Collected Papers of Albert Einstein,* trans. Anna Beck (Princeton, NJ: Princeton University Press, 1989), vol. 2, p. 87.
8. Reference to his paper and M. J. Klein, "The First Phase of the Bohr-Einstein Dialogue," in Russell McCormmach, ed., *Historical Studies in the Physical Sciences* (Philadelphia: University of Pennsylvania Press, 1970), vol. 2, pp. 1–39.
9. A. Einstein, *Physikalische Zeitschrift,* vol. 10 (1909), p. 817.
10. Klein, "Bohr-Einstein Dialogue," p. 7.

11. Franz von Baader, "Über den Blitz als Vater des Lichts," in Franz von Hoffmann, ed., *Sämtliche Werke* (Aalen, Ger.: Scientia, 1963), vol. II.

12. Antoine Faivre, "Ténèbre, Éclair et Lumière chez Franz von Baader," *Lumière et Cosmos* (Paris: Albin Michel, 1981), p. 268.

13. Robert H. Eather, *Majestic Lights* (Washington, D.C.: American Geophysical Union, 1980); and Syun-ichi Akasofu, "The Aurora," in *Light from the Sky* (San Francisco: W. H. Freeman, 1980).

14. Werner Heisenberg, *Encounters with Einstein* (Princeton, NJ: Princeton University Press, 1989), pp. 112ff.

15. Einstein, quoted in Abraham Pais, *Subtle Is the Lord* (New York: Oxford University Press, 1982), pp. 416–17.

16. It is called the semi-classical or neo-classical theory of quantum mechanics.

17. Ernst Blass, "The Old Café des Westens," in *The Era of Expressionism*, ed. Paul Raabe, trans. J. M. Ritchie (Dallas, TX: Riverrun Press, 1980), p. 29.

18. Gustav Mie, inaugural lecture, January 26, 1925, University of Freiburg in Breslau, quoted by Paul Forman, "Weimar Culture, Causality and Quantum Theory (1918–27)" in *Historical Studies in the Physical Sciences*, ed. Russell McCormmach (Philadelphia: University of Pennsylvania Press, 1971), vol. 3, pp. 1–116."

19. Werner Haftmann, *The Mind and Work of Paul Klee* (London: Faber & Faber, 1954), pp. 57–58.

Chapter 10 / Of Relativity and the Beautiful

1. Albert Einstein, *Ideas and Opinions,* "On the Method of Theoretical Physics" (New York: Crown Publishers, 1954), pp. 270–76.
2. Quoted in Abraham Pais, *Subtle Is the Lord,* p. 131.
3. Henri Poincaré, quoted in Jeremy Bernstein, *Einstein* (New York: Viking, 1973), p. 103.
4. A. Einstein, in Paul Arthur Schlipp, ed., *Albert Einstein: Philosopher-Scientist* (La Salle, IL: Open Court, 1949), p. 25.
5. Einstein, "Ether and the Theory of Relativity," a 1920 address in *Sidelights on Relativity* (New York: Dover, 1983).
6. See Arthur Miller, *Albert Einstein's Special Theory of Relativity* (Reading, MA: Addison-Wesley Publishing Co., 1981), p. 166.
7. On the question of causality and relativity, see David Bohm, *The Special Theory of Relativity* (New York: W. A. Benjamin, 1965), chap. 28.
8. There are special combinations of E and B that *are* invariant. See J. D. Jackson, *Classical Electrodynamics* (New York: Wiley, 1975), pp. 517–19.
9. David C. Lindberg, "Medieval Latin Theories of the Speed of Light," in *Roemer et la Vitesse de la Lumière* (Paris: Vrin, 1978), pp. 45–72.
10. Aristotle, *De anima,* II, 7; *De sensu,* VI. See Sabra, *Theories of Light,* chap. 2.
11. Lindberg, *Theories of Vision,* p. 48.
12. Sabra, *Theories of Light,* p. 48.
13. M.-A. Tonnelat, "Vitesse de la Lumière et Relativité," in *Roemer et la Vitesse de la Lumière,* p. 282.
14. Albert Einstein, "On the Electrodynamics of Moving Bodies," trans. Arthur Miller, in *Albert Einstein's Special Theory of Relativity,* p. 401.

15. James H. Smith, *Introduction to Special Relativity* (Champaign, IL: Stipes Publishing, 1965), chap. 2.
16. A. Brillet and J. Hall, "Improved Laser Test of the Isotropy of Space," *Physical Review Letters*, vol. 42 (1979), pp. 549–52. See also O'Hanian's *Classical Electrodynamics* (Boston: Allyn and Bacon, 1988), pp. 157–164.
17. D. Newman, G. W. Ford, A. Rich, and E. Sweetman, "Precision Experimental Verification of Special Relativity," *Physical Review Letters*, vol. 40 (1978), pp. 1355–58.
18. See J. Terrel, *Physical Review*, vol. 116 (1959), pp. 1041ff.
19. Goethe, *Scientific Studies*, p. 263.
20. See John Lobell, *Between Silence and Light: Spirit in the Architecture of Louis I. Kahn* (Boulder, CO: Shambhala, 1979); and Louis Kahn, *Light Is the Theme*, comments on architecture complied by Nell E. Johnson (Fort Worth, TX: Kimbell Art Foundation, 1975).
21. Pais, *Subtle Is the Lord*, chap. 9.
22. Einstein, "Ether and Relativity."
23. Quoted in Emil Wolf, "Einstein's Researches on the Nature of Light," *Optics News*, vol. 5, no. 1 (Winter 1979), pp. 24–39.
24. Richard Feynman, *Lectures on Physics* (Reading, MA: Addison-Wesley, 1968), vol. II, chap. 19, p. 9.
25. "Metaphysical Derivations of a Law of Refraction: Damianos and Grosseteste," Bruce S. Eastwood, *Archive for the History of Exact Sciences*, vol. 6 (1970), pp. 224–36; and *Journal of the History of Ideas*, vol. 28 (1967), pp. 403–14.
26. Plato, *Timaeus*, 46e.
27. For a sober and detailed treatment, see Wolfgang Yourgrau and Stanley Mandelstam, *Variational Principles in Dynamics and Quantum Theory*, 3rd edition (London: Sir Isaac Pitman & Sons, 1968).
28. Einstein, *Ideas and Opinions*, p. 228.
29. Friedrich Schiller, *The Aesthetic Letters*, trans. J. Weiss (Boston: C. C. Little and J. Brown, 1845).

Chapter 11 / Least Light:
A Contemporary View

1. P. Grangier, G. Roger, and A. Aspect, "Experimental Evidence for a Photon Anticorrelation Effect on a Beamsplitter: A New Light on Single Photon Interferences," in *Europhysics Letters*, vol. 1 (January 1986). On the general development of the modern conception of the photon, see Richard Kidd, James Ardini, and Anatol Anton, *American Journal of Physics*, vol. 56 (1988), pp. 27–35.

2. Quoted in Arthur Fine, *The Shaky Game* (Chicago: University of Chicago Press, 1986), p. 106.

3. N. Bohr, "Discussions with Einstein on Epistemological Problems in Atomic Physics," in *Albert Einstein: Philosopher-Scientist*, P. A. Schlipp, ed. (La Salle, IL: Open Court, 1949), pp. 200–41.

4. J. A. Wheeler, in *The Ghost and the Atom*, eds. P. C. W. Davies and J. R. Brown (Cambridge: Cambridge University Press, 1986), p. 64; and *Mathematical Foundations of Quantum Theory*, ed. A. R. Marlow (New York: Academic Press, 1978); and C. F. von Weizsäcker, *Zeitschrift für Physik*, vol. 70 (1931), p. 114.

5. T. Hellmuth, H. Walther, A. Zajonc, and W. Schleich, "Delayed-Choice Experiments in Quantum Interference," *Physical Review*, vol. 35 (1987), pp. 2532–41. See also John Horgan, "Quantum Philosophy," *Scientific American* (July 1992), pp. 94–104.

6. Wheeler, *The Ghost in the Atom*, p. 69.

7. Fritz Rohrlich, *From Paradox to Reality* (Cambridge: Cambridge University Press, 1987), p. 22.

8. David Bohm, *Wholeness and the Implicate Order* (New York: Ark Paperbacks, 1987).

9. J. S. Bell, *Speakable and Unspeakable in Quantum Mechanics* (Cambridge: Cambridge University Press, 1987), p. 27.

10. A. Aspect, et al., "Experimental realization of Einstein-

Podolsky-Rosen-Bohm *Gedankenexperiment:* A new niolation of Bell's inequalities," *Physical Review Letters,* vol. 49 (1982), pp. 91–94; and "Experimental test of Bell's inequalities using time-varying analyzers," pp. 1804–7.

11. These are the so-called quantum beat experiments.
12. Erwin Schrödinger, *What Is Life? and Other Scientific Essays* (Garden City, NY: Doubleday, 1956), pp. 161–62.
13. U. Bonse and H. Rauch, eds., *Neutron Interferometer* (New York: Oxford University Press, 1979).
14. See "Making Waves with Interfering Atoms," *Science,* vol. 252 (May 17, 1991), pp. 921–22.
15. See articles by C. Tesche, S. Washburn, and R. Webb, for example, in *New Techniques and Ideas in Quantum Measurement Theory,* ed. Daniel Greenberger, vol. 480 (1986), *Annals of the New York Academy of Sciences,* pt. II.
16. D. Bohm, B. J. Hiley, and P. N. Kaloyerou, "An Ontological Basis for the Quantum Theory," *Physics Reports,* vol. 144, no. 6 (1987), pp. 321–75.
17. Francis Thompson, quoted in Alan J. Friedman and Carol C. Dorley, *Einstein as Myth and Muse* (Cambridge: Cambridge University Press, 1985), p. 44.
18. Hesiod, *Theogony.*
19. Aristotle, *Physics,* trans. Richard Hope (Lincoln, NE: University of Nebraska Press, 1961), p. 58.
20. T. D. Newton and E. P. Wigner, *Reviews of Modern Physics,* vol. 21 (1949), pp. 400–06; see also E. R. Pike and Sarben Sarkar, "Photons and Interference," in E. R. Pike and Sarben Sarkar, eds., *Frontiers in Quantum Optics* (Boston: Adam Hilger, 1986), pp. 282–317.
21. Murray Sargent III, Marlan Scully, and Willis E. Lamb, Jr., *Laser Physics* (Reading, MA: Addison-Wesley Publishing Co., 1974), p. 228.
22. J. A. Valdmanis and N. H. Abramson, "Holographic Imaging Captures Light in Flight," *Laser Focus World,* February 1991, pp. 111–17.

23. James Turrell, in *Occluded Front*, ed. Julia Brown (Los Angeles: The Lapis Press, 1985), p. 46.
24. Gerard de Nerval, *Le Christ aux Oliviers* in *Dictionary of Foreign Quotations*, compiled by Robert and Mary Collison (New York: Facts on File, 1980), p. 138.
25. H. G. B. Casimir and D. Polder, "The Influence of Retardation on the London-van der Waals Forces," *Physical Review*, vol. 73 (1948), p. 360.
26. H. B. G. Casimir, "On the attraction between two perfectly conducting plates," *Proceeding of the Koninklijke Nederlandse Akademie van Wetenschappen*, vol. 51 (1948), p. 793.
27. See I. J. R. Aitchison, "Nothing's plenty: The vacuum in modern quantum field theory," *Contemporary Physics*, vol. 26 (1985).
28. Victor Hugo, *Dernière Gerbe*, November 26, 1876.
29. Herder, speaking of life, quoted in Bachelard, *The Flame of a Candle*, p. 2.

Chapter 12 / Seeing Light

1. See Margaret Miles, "Vision," *The Journal of Religion*, vol. 63 (1984), pp. 125–42.
2. Dante Alighieri, *The Paradiso*, trans. John Ciardi (New York: New American Library, 1970), canto 31, lines 22ff.
3. St. Bernard (or an imitator), quoted in Millard Miess, "Light as Form and Symbol in Some Fifteenth-Century Paintings," *Art Bulletin*, vol. 27 (1945), pp. 175–81.
4. Otto von Simson, *The Gothic Cathedral* (Princeton, NJ: Princeton University Press, 1988), 3rd ed., p. 55; and all of chap. 2.
5. Ernest Gellner, *Plough, Sword and Book: The Structure of Human History* (Chicago: The University of Chicago Press, 1988), chap. 2.

6. Hesiod, *Works and Days*, pp. 110ff.

7. Vico, quoted in Charles M. Radding, *A World Made by Men: Cognition and Society, 400–1200* (Chapel Hill, NC: University of North Carolina Press, 1985).

8. Louis L. Noble, *The Course of Empire* (New York: Lamport, Blakeman & Law, 1853), p. 289.

9. See plate 5 of Elizabeth Sears, *The Ages of Man: Medieval Interpretations of the Life Cycle* (Princeton, NJ: Princeton University Press, 1986).

10. Novalis, *The Disciples at Sais and Other Fragments*, trans. F.V.M.T. and U.C.B. (London: Methuen and Co., 1903), pp. 137.

11. *Paul Cézanne Letters*, ed. John Rewald (New York: Hacker Art Books, 1976), p. 327, to his son, September 8, 1906.

12. Cézanne, *Letters*, pp. 303 and 306, July 25, 1904.

13. Buddhaghosa, *Path of Purity*, trans. Pe Maung Tin (Pali Text Society, 1975), p. 145.

14. Buddhaghosa, *Path of Purity*, p. 200.

15. J. W. Goethe, *Goethes Gespräche* (Leipzig, Ger.: Biedermann, 1901–11), vol. 2, p. 245.

16. Goethe, *Goethes Werke, Hamburger Ausgabe*, vol. 13, p. 323.

17. Goethe, *Goethes Werke, Hamburger Ausgabe*, vol. 13, p. 38.

Acknowledgments

The writing of this book would not have been possible without the help of many friends and teachers. My interest in Goethe, Steiner, and the humanistic and spiritual dimensions to light were first piqued by Professors Ernst Katz and Alan Cottrell, two early mentors, to whom I owe a great debt. My study of light from the vantage point of quantum physics was stimulated by research visits to the École Normale Supériure with Marie Anne Bouchiat, the Max Planck Institute for Quantum Optics with Herbert Walther and Marlan Scully, the University of Hanover with Jürgen Mlynek, and the University of Rochester with Leonard Mandel. Among many local colleagues whose conversations provoked and clarified my thinking I would like to give special mention to George Greenstein, Herbert Bernstein, K. Jaganathan, Larry Hunter, Dudley Towne, and Bob Krotkov. For literary and artistic themes, I thank Frederick Amrine, Douglas Patey, Douglas Miller, Christopher Bamford, and Joel Upton for their interest. I would like to thank William Irwin Thompson for including me in Lindisfarne gatherings, which provided the occasion for developing certain threads of the book, and also his son Evan Thompson for pushing me to think ever more deeply about vision.

In addition to the many colleagues important to the creation of this book, I would like to thank Amherst College, especially its library staff, and Laurence Rockefeller for support during a sabbatical leave. Without such support, writing would have been even more protracted than it already was. Of course, without the encouragement of my family, and the excitement of my sons, August and Tristan, the project would have been impossible.

Finally, my thanks to Leslie Meredith for her early enthusiasm and careful editing, and most especially to Patricia van der Leun whose constant support, good judgment, and unflagging attention to the book helped shape it from inception through publication.

To all of these goes much of the credit for what is good in *Catching the Light*. I reserve whatever errors and infelicities it contains to myself.

Index

Grateful acknowledgment is made for permission to reprint from the following:

Page 33: The Bodleian Library, University of Oxford, Vet. B3.e.105., p. 125.

Page 60: Theories of Vision from al-Kindi to Kepler, by David C. Lindberg, edited by Allen G. Debus, copyright © 1976 The University of Chicago Press.

Page 62: Duccio, *The Temptation of Christ on the Mountain,* copyright The Frick Collection, New York.

Page 64: From "Pictorial Perception and Culture," by Jan B. Deregowski. Copyright © 1972 by Scientific American, Inc. All rights reserved.

Page 66: Reprinted from *The Painter's Manual,* by Albrecht Dürer. Translated by Walter L. Strauss (Abaris Books, 1977), fig. 67, p. 434.

Page 68: Charles D. O'Malley and J. B. de C. M. Saunders, *Leonardo da Vinci on the Human Body,* copyright 1952 by Henry Schuman, Inc., New York.

Page 72: Alinari/Art Resource, New York.

Page 108: Photo courtesy M. Cagnet, M. Francon, and J. C. Thrierr: *Atlas optischer Erscheinungen,* Berlin–Heidelberg–New York: Springer, 1962.

Page 118: Fundamentals of Physics, 3d ed., by David Halliday and Robert Resnick, copyright © 1988 by John Wiley & Sons, reprinted by permission of John Wiley & Sons, Inc.

Page 131: Figure from *Physics: For Scientists and Engineers,* Third Edition, copyright © 1990 by Raymond A. Serway, reprinted by permission of Saunders College Publishing.

Page 141: Faust in His Study, etching by Rembrandt, B. 270. Reprinted by permission of Rijksmuseum-Stichting.

Credits